Sound Pressure

Media Philosophy

Series Editors:
Eleni Ikoniadou, Senior Tutor (Research) in Visual Communication at the Royal College of Art
Scott Wilson, Professor of Cultural Theory at the London Graduate School and the School of Performance and Screen Studies, Kingston University

The Media Philosophy series seeks to transform thinking about media by inciting a turn towards accounting for their autonomy and 'eventness', for machine agency, and for the new modalities of thought and experience that they enable. The series showcases the 'transcontinental' work of established and emerging thinkers whose work engages with questions about the reshuffling of subjectivity, of temporality, of perceptions and of relations vis-à-vis computation, automation and digitalisation as the current twenty-first-century conditions of life and thought. The books in this series understand media as a vehicle for transformation, as affective, unpredictable and nonlinear, and move past its consistent misconception as pure matter-of-fact actuality.

For Media Philosophy, it is not simply a question of bringing philosophy to bear on an area usually considered an object of sociological or historical concern but of looking at how developments in media and technology pose profound questions for philosophy and conceptions of knowledge, being, intelligence, information, the body, aesthetics, war, death. At the same time, media and philosophy are not viewed as reducible to each other's internal concerns and constraints, and thus it is never merely a matter of formulating a philosophy of the media; rather the series creates a space for the reciprocal contagion of ideas between the disciplines and the generation of new mutations from their transversals. With their affects cutting across creative processes, ethicoaesthetic experimentations and biotechnological assemblages, the unfolding media events of our age provide different points of intervention for thought, necessarily embedded as ever in the medium of its technical support, to continually reinvent itself and the world.

'The new automatism is worthless in itself if it is not put to the service of a powerful, obscure, condensed will to art, aspiring to deploy itself through involuntary movements which nonetheless do not restrict it'.

—Eleni Ikoniadou and Scott Wilson

Software Theory: A Cultural and Philosophical Study, by Federica Frabetti
Media after Kittler, edited by Eleni Ikoniadou and Scott Wilson
Chronopoetics: The Temporal Being and Operativity of Technological Media, by Wolfgang Ernst, translated by Anthony Enns
The Changing Face of Alterity: Communication, Technology and Other Subjects, edited by David J. Gunkel, Ciro Marcondes Filho and Dieter Mersch
Technotopia: A Media Genealogy of Net Cultures, by Clemens Apprich, translated by Aileen Derieg
Contingent Computation: Abstraction, Experience, and Indeterminacy in Computational Aesthetics, by M. Beatrice Fazi
Recursivity and Contingency, by Yuk Hui
Sound Pressure: How Speaker Systems Influence, Manipulate and Torture, by Toby Heys
Media Arts in the XXI Century: Archaeologies, Theories, Preservation, by Valentino Catricalà, translated by Arabella Ciampi (forthcoming)

Sound Pressure

How Speaker Systems Influence, Manipulate and Torture

Toby Heys

London • New York

Published by Rowman & Littlefield International Ltd.
6 Tinworth Street, London, SE11 5AL
www.rowmaninternational.com

Rowman & Littlefield International Ltd. is an affiliate of Rowman & Littlefield
4501 Forbes Boulevard, Suite 200, Lanham, Maryland 20706, USA
With additional offices in Boulder, New York, Toronto (Canada), and Plymouth (UK)
www.rowman.com

Copyright © 2019 by Toby Heys

All rights reserved. No part of this book may be reproduced in any form or by any electronic or mechanical means, including information-storage and -retrieval systems, without written permission from the publisher, except by a reviewer who may quote passages in a review.

British Library Cataloguing in Publication Data

A catalogue record for this book is available from the British Library.

ISBN: HB 978-1-78661-112-3

Library of Congress Cataloging-in-Publication Data Is Available

ISBN 978-1-78661-112-3 (cloth)
ISBN 978-1-5381-4794-8 (pbk)
ISBN 978-1-78661-113-0 (electronic)

Contents

Foreword: Speakers *Dave Tompkins*	ix
Acknowledgements	xv
Introduction: Frequency-Based Force	**1**
Oscillating Perception	3
Operations of the Antenna-Body	4
Frankenstein in Frequencies	4
Propositions	7
Sonic Geography	9
Sound as Virus	13
Waveformed Language and Listening	15
1 Muzak's Influence in the Fordist Factory	**23**
Distributed Rhythms of Labour	23
Audioanalgesia	27
Transnational Harmonies	31
Masters of Discipline	32
Mass Neural Network	35
Disconnecting the Global Village	41
A Fleshy Cadence of the Antenna-Body	44
2 Surround-Sound Manipulation at the Waco Siege	**49**
Body of Noise/Body in Noise	49
Channelling Violence	55
A Psychological Symphony of Disaster	58
The Fifty-One-Day Diary of a Failed Rock Star	60
Virasonic Attack	63

	Inverting Wilderness	66
	The Acoustic Ontology of an Outsider	68
	Audiotopia	70
	Loudhailers of the Living Dead	72
3	**Torture in Black Ecstasy at Guantánamo Bay**	**77**
	Sonic Intensity	77
	Disco Inferno	82
	Asymmetric Conflict	87
	Break It Up, Break It Up, Break It Up, Break Down	89
	Excavation, Autopsy, Exorcism	91
	Reflecting Silence	94
	Violence and the Voice	97
	The Instrumentality of Sonic Pain	100
	Fear of Music	102
	Futility Music	103
	The Repetitive Cell	107
	Four Walls of Sound	110
	The Excessive Reality of Waveformed Pressure	112
	Cultural Compression	115
	Worn-Out Phantom Limbs	117
4	**The Covert Aims of Directional Ultrasound**	**123**
	Heterodyning Logic	123
	HyperSonic Sound	124
	Audio Spotlight	124
	Ultrasonic Utility	126
	Infrasonic Utility	127
	The Nonsound of Nonlethal Weapons	128
	Music against Youth	131
	Movements of an Unheard Body	132
	A Whispering Parasite and Siamese Consciousness	133
	Changing the Channels of Duplicity	134
	The Psychogeography of a Schizophrenic's Ear	137
	A Silent Killing in the Transauditive Era	139
	We Are Never outside of Geography	141
	Inaudible Infection	144
	Engineering the Twenty-First-Century Ear	146
5	**Whispering to Talking Windows**	**151**
	Feonic: Resequencing Materiality	151
	Furniture Music	153
	Music as Competition	156

Teflon Acoustics	157
Bone Logic	164
Haunted by Capacity	168
Stone Tape Theories of Animism	170
Windows on the Undead Web	172
Conclusion: Phantom Sound Systems	**177**
Acoustic Attack: Cuba	177
Acoustic Attack: China	179
Subvocalisations of the AlterEgo	180
Pressure as Mimetic Ordnance	183
References	189
Index	207
About the Authors	215

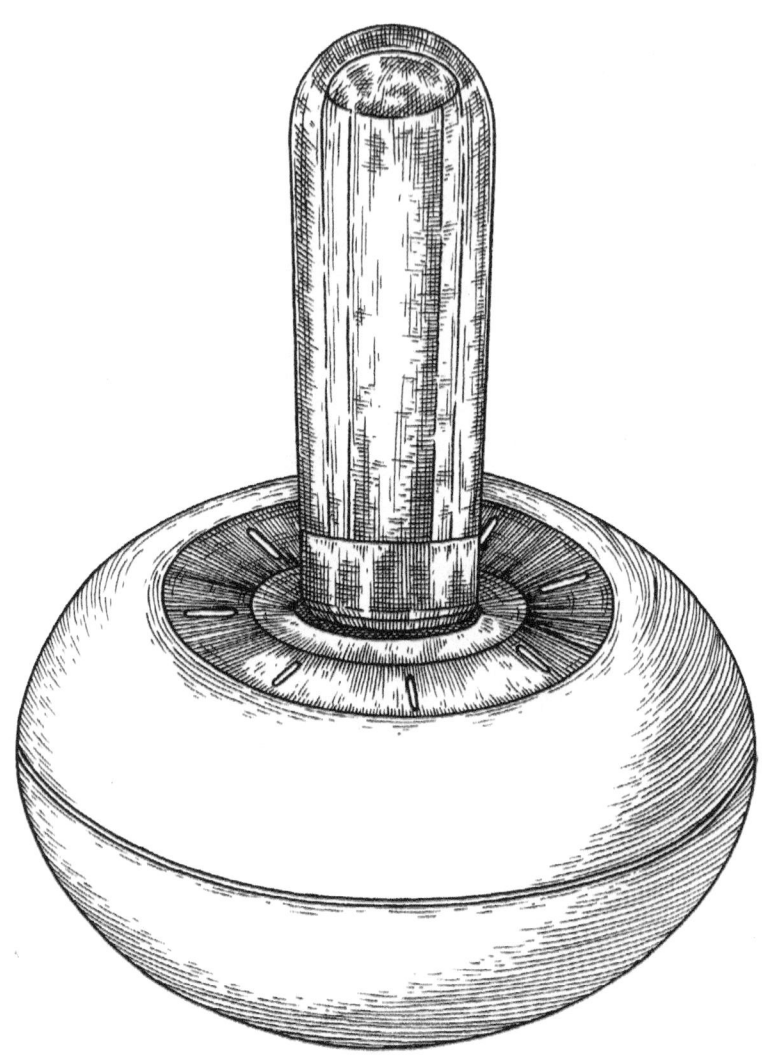

I-MU Vibro Vibration Speaker. *Illustration by Krystian Griffiths*

Foreword: Speakers

Dave Tompkins

Speakers have assimilated into backdrop, the natural Amazon Echo environment. Walls, counters, tables and windows are no longer just reactionary landscapes but flat, surface-to-air carriers of vibration, or as Toby Heys puts it 'molecular disturbances'. Convenient, dustable, compatible, even friendly. A voice sells the room by granting it intelligence so residents can 'join the conversation' between sleek inanimate objects. The speaker as a voice inhabits speaker as music component, but identified as neither, just a 'virtual assistant', so-called across the room, no longer a cabinet with a conical pulse and a baffle but still part of a larger system. Asymmetrical, all-in-one! A play at omniscience without knowing its L from its R[1]. The sound-system projection of self, larger than life, once expanding circumference, eighteens, twenty-twos, rippling margins and diaspora, has contracted into a three-inch cylinder with 'beamforming technology'. 'Fill any room'!

Alexa, play GrandMixer DXT's memory of dawn in the West Bronx in 1976. Play 'Heaven and Hell Is on Earth'. Clap for space.

DXT[2], then a teenaged Derek Showard, sat on the hood of a charred sedan in its final parking space at 167th and Grant Avenue, on a block abandoned by the city itself more than by its tenants. DJ Smokey had returned home after his crew, the Smokatrons, had been decimated by Kool Herc in a battle at the Webster Avenue PAL center. Drowned out, Smokey's wall of speakers had appeared to stand in silence, puckering, 'miming'[3]. (Herc's massive speakers had once been filmed riding in the backseat of his convertible like hi-rise dignitaries at a fat-cat recline, transported on the business of decibel annihilation.)

Smokey's block returned DXT to some apocalyptic Omega Man rubble he'd caught on television. As a child, he'd 'put things in his head' in an attempt to block out the human suffering witnessed in his Bronx neighbourhood. He would imagine the retrofuture technology from science fiction on TV transforming into stereo equipment. The Jupiter 2 platter as a turntable. Rocket boosters as speakers. Long before DXT appeared on TV himself at the 1984 Grammys, performing in front of James Brown and Michael Jackson with a turntable hooked to a guitar flange, the kid rewired his own mind with acoustic configurations that remapped his living room and filled the window looking out at Edenwald Houses off East 233rd Street. The imagined speaker became a belief system.[4]

DXT's mentors and peers put speakers in windows, back turned to the room, projecting outward, a shared recurrent memory of black and Latinx childhoods, circulated, generously, by those who lived it. Like Norman Bottom, the 'knockout artist' who guarded DXT's records, many of which bore Smokey's signature. Was 'Heaven and Hell' among them? The cabinet standing on DJ Smokey's stoop at sun-up rumbled with the 20th Century Steel Band from Trinidad, reviving the set that Herc had drowned out earlier that night[5]. Though you couldn't hear Smokey at the PAL battle, his records could've been sensed in the sub range, hairs on low end. The West Indies sense of Third Ear[6], the diaphragm, vibrating body as speaker, a gut feeling that would send kids like Derek Showard over to Smokey's place later. They sat on Pontiac husks listening to 20th Century Steel, the chorus riding that stretch-hell note. Just the chant and echo claps at first, mixed in with the creak and sway of empty buildings[7]. Shadows of blown windows, playing the walls.

The early subway commute from the Bronx calls for a special mobile sound system and a nap. Earbuds and file compression are components of the imagined speaker, sacrificing bass for mobility while ironically excluding the rest of the body from the listening experience, leaving it behind like a depressed 'Super Bass' button in the Walkman past[8]. It's all in your head, nodding off against smeared glass, brainwaves cycling at ten hertz. The speaker vanishes into Talking Windows, bypassing both stereo construct and tympanic membrane. Traditionally, the window has played a rattling, tinted rearview part in the mobile sound system, triggered by the King Kong in the trunk while announcing its future perception, in a moment crawling by[9]. Talking Windows are flattened speakers, as if stomped by Maggotron's fifty-foot stocky 'Quadzilla' on The Bass That Ate Miami album cover. This acoustically treated Feonic glass transfers sound directly into our heads via bone conduction, a venerable clampdown technique that left Beethoven's tooth prints in a rod but allowed him to hear his own Fifth. The direct approach.

Invisible, lacking transparency and ideal for advertising. It gives the impression that a plug for Amazon is actually your receptive conscious, as if the power of agency was yours all along. But that tingle of excitement is just the skull vibrating. Talking Windows are brought to you, in your commuter nap, by Terfenol-D, an alloy devised by the Naval Ordnance Laboratory in the 1970s. The D is for dysprosium, a rare-earth metal that slipped between the headlines of Trump's trade war on China in 2018. Periodic tables were checked for tariffs. Atomic number 66, Dy, with a silver disco luster. Talking Windows could deliver new headlines with old anxieties under its breath. The Nuclear Test Ban Treaty with Russia has been suspended, but dysprosium rods can help keep the reactors cool. The navy found Terfenol-D applications for sonar and 'sub-bottom' sediment mapping, while a Department of Energy lab in Iowa discovered the alloy could unstink manure—the same lab that accused grad students from China of stealing its Terfenol-D secrets. All your paranoid rare-earth metal news delivered through the reflection of your vibrating familiar, half-realized at the next stop.

Talking Windows could also compete for market-driven mental space with the 'imagine speaker', a voice that's been in our heads ever since we first learned to read. Subvocalisation occurs when silently communicating words on the page or screen to the brain. The speech mechanism working in silence. Speed readers suggest humming while reading to disable the imagine speaker. But who, or what, shuts off the hum?

The so-called acoustic attacks on American embassies in Cuba and China have appeared in quotes as 'part of the conversation' created by subvocalised readings as the story germinated through media outlets, disembodied comments and waves of takes. Imaginespeak begat an imagined speaker with a mind of its own. A sound system for the ages! Among the symptoms reported from embassies in Cuba and China were brain trauma. In January 2019, Nature published a neuroscience study posing the idea that a vocoder could be used to interpret our thoughts (Akbari, Khalighinejad, Herrero, Mehta, and Mesgarani 2019). Working with epilepsy patients during neurosurgery, the Mind Brain Behavior Institute at Columbia University reported that brainwaves could 'train' a vocoder algorithm to translate signals into intelligible speech, aided by a deep neural AI mimic and subvocal signaling.

'Experts, trying to record and decode these patterns, see a future in which thoughts need not remain hidden inside the brain—but instead could be translated into verbal speech at will' (Zuckerman Institute at Columbia University 2019). It would be a potential breakthrough in neuroprosthetics, allowing ALS patients or stroke victims to speak, whether through a laptop or Amazon Echo[10]. A vocoder-to-vocoder sound system activated by thought, speech urges harvested from a neural implant and transmitted to surface level speakers, in the voice of a grocery list and directions. Can they recognize each other? The imagined speaker network serves as an Acousmonium, its drivers connected through desire. This is sound-system wall nostalgia compressed and flattened into convenience, a daily reminder of the perceived need to metabolise the excess and perhaps, on occasion, listen to 'Heaven and Hell Is on Earth'. It is a projection of wishful thinking.

NOTES

1. Don't let the symmetry fool you: I once saw a pair of bookshelf speakers that turned out to be air purifiers.
2. Originally DST, he changed his name to DXT after his brother was killed, drawing inspiration from Malcolm X and his favourite film growing up, *THX 1138*.
3. This section is based on interviews with DXT by Jeff Mao (*RBMA Daily*, 2018) and Dave Tompkins (*Wax Poetics*, 2014, and *How to Wreck a Nice Beach*, 2010).
4. As resident DJ at The Roxy, DXT often backed Rammellzee, who later wore speakers on his body, encoded into his alphabet armor.
5. '"Turn . . . turn . . . your system . . . system . . . down . . . down!" And it's cutting through, it sounds like he's on Smoke's system. It's cutting through everything that he's playing. And Clark goes, "Or we will *drown you . . . drown you . . . drown you!*"' DXT's impression of Clark Kent on the Herculoids' echo chamber, as told (told) to (to) Jeff (Jeff) Mao (Mao) for the Jalal Mansur Nuriddin obituary (Mao 2018); emphasis original.
6. Foreword author's interview with Carlos Malcolm. Also see *Sonic Bodies* by Julian Henriques (2011).
7. 'Errrt, errrt, errrrt . . .' DXT to Mao (2018).
8. Did that shit actually work?
9. Lil Wayne, 'King Kong' from *Da Drought 3*, surviving a drowned world and a broken system. 'If you hear me 'fore you see me . . .'
10. Or the marketed Echo 2nd Generation: When asked 'Are you a vocoder?', Alexa responded, 'Hmm, I'm not sure'. Alexa, however, was able to play 'I Am a Vocoder' by Cat Gay Park.

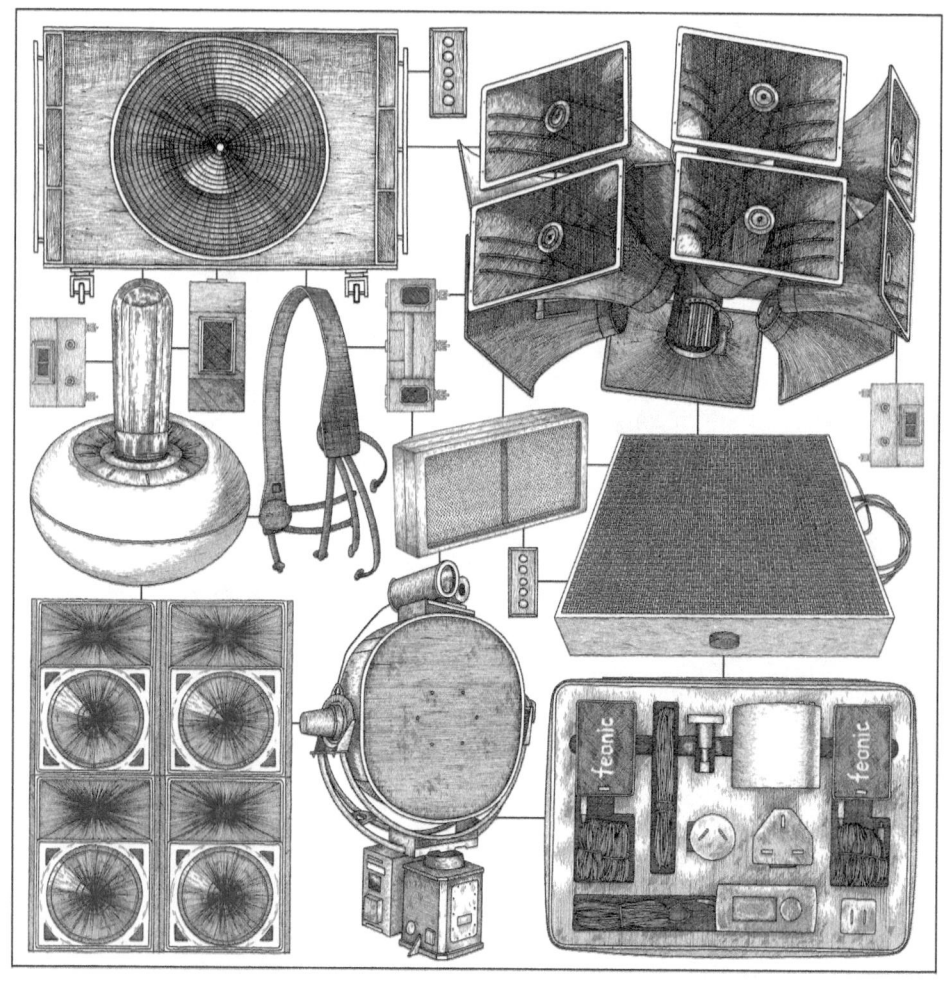

Distributed Speaker System. *Illustration by Krystian Griffiths*

Acknowledgements

Caro, Charly and Ivo, thank you for putting up with my static long nights.

Massive gratitude for the harsh editing of Robin Mackay at Urbanomic and Marsha Courneya, wonderful illustrations provided by Krystian Griffiths, inspired foreword by Dave Tompkins.

Thanks to my fellow waveformed thinkers for your conversations, advice and suggestions over the past decade or so—Robert Saucier, Kristen Gallerneaux, Colin Fallows, Lewis Braden, Conor Thomas, Lee Gamble, John Hyatt, Andrew Coleman, Thomas Couzinier, Nik Nowak, Jonathan Sterne, Alastair Cameron, Naut Humon.

In appreciation of the continued insights of David Jackson, Richard Wistreich, Clive Parkinson, Hayley Walsh, Jim Aulich, Lance Weiler and all at Digital Storytelling Lab, Annetta Kapon, Berengere Marin-Dubuard, Berthold Schoene, Marcio Lana-Lopez.

Thanks to all at AUDINT—Steve Goodman, Souzanna Zamfe, Eleni Ikoniadou, Patrick Defasten and IREX[2].

And finally to the global waveformed network that this book has briefly plugged into and traced a line in and out of. To the distributed speaker system that consists of Curdlers, Technomads, Soundlazers, Intonarumori, woofers, HyperSonic Sound Systems, passive monitors, Phonetrons, sound bars, Long Range Acoustic Devices, Mosquitos, Medusas, public-address systems, HyperSpikes, horns, sonic cannons, Audaunters, Alter/Ego, Talking Windows, Active Denial Systems, rotary woofers, Audio Spotlights, Vitaphones, Aversive Audible Acoustic Devices, Active Monitors, tweeters, M-Forces, electrostatic flat panels, ribbon drivers, multi-cell flat diaphragms, Whispering Windows, plasma arcs, piezoelectrics, ring-vortex cannons, bullhorns, Thorophones, moving-coils, SoundCommanders, Outacoms, subwoofers, planar magnetics, Sequential Arc Discharge Acoustic Generators.

LRAD 1000RX. *Illustration by Krystian Griffiths*

Introduction
Frequency-Based Force

Speaker systems exert pressure on us physically and psychologically. So long as the transmission of the infrasonic (below the hearing range), sonic and ultrasonic (above the hearing range) frequencies that produce this force are perceived as beneficial, we tend to disregard such dynamics. The production of entertainment and pleasure, the dissemination of public information and medical checks are a few of the more obvious ways in which we associate domestic and public frequency-based technologies and the content they emit, with a positive instrumentality. Taking this one step further, we utilise speakers and sound systems on a daily basis to modulate our emotional states, to alert us to life-threatening dangers such as fire and to reveal the internal workings of the body. But what happens when such technologies are called into service for other, more insidious, purposes? When waveformed pressure is built up and released at the behest of agendas of affective influence, programmed manipulation and, ultimately, political torture?

Sound Pressure is a transdisciplinary and transhistorical investigation into civilian and battlefield contexts in which speaker systems have been utilised by the military-industrial and military-entertainment complexes to apply pressure to the collective and individuated body. One of the core propositions of the book is that a line can be traced from the inception of wired radio speaker systems[1] and their introduction into US industrial factories in 1922 (chapter 1) through the development of sonic strategies[2] based primarily on the scoring of architectonic spatiality, cycles of repetition and the enveloping dynamics of surround sound to the sonic torture perpetrated in Guantánamo Bay during the first decade of the twenty-first century (chapter 3). The FBI's use of an extended sound system to envelop and harass members of the Branch Davidians within their compound during the

Waco siege of 1993 (chapter 2) mediates between these episodes. Joining the dots of the historical deployment of acoustic force, this chapter also supports the contention that the acoustic techniques utilised in Guantánamo's torture cells represent the final modality and the logical conclusion of these strategies, which have evolved between civilian and military contexts over the past eighty years.

The instrumentalising of speaker systems that occurs subsequently, post-Guantánamo, prompts the second core premise of the book. It is contended that covert speaker technologies come to symbolise a shift in episteme in regard to the application of waveformed pressure, as the dynamics of directional ultrasound technology—heralded by the HyperSonic Sound System (HSS)—signal the orchestration of a new set of frequency-based relations between transmitter and receiver, speaker system and architectural context and civilian and war-torn environment (chapter 4). This dynamic equation of subject, space and waveformed emission shifts again as the recently developed flat-panel speaker system's potential to turn any and all level surfaces into speakers ambiently weaponises the notion of surround sound (chapter 5). By inferring that there is an as-yet-unperceived connectedness to the metanetwork of speakers within which we exist, this final section opens up a speculative channel of enquiry into notions of waveformed sentience, questioning the pressures that may come to bear on a future AI (an audio, rather than artificial, intelligence).

As waveformed strategies become more abstracted, verging on the science fictional as in the case of the acoustic attacks on members of staff at the American embassy in Havana, Cuba, in 2017, the virtual nebulosity of both their technical possibilities and their online existence as speculative memes comes to reflect the ideational spatiality within which unsound (ultrasound and infrasound) operates[3]. To speak about pressure in the physical sciences is to articulate the ways in which force is applied to and distributed across the surface of an object. But when we speak of frequency-based pressure in relation to the materiality of the distributed body—implying the interweaving of a psychology, a physiology and a virtual data twin—we are dealing with a more porous and complex equation of force. Each of this book's five chapters presents a case study of a particular sound system along with its attendant techniques, agendas and spatialities in order to question how perceptions of duress, compression and provocation are coming to redefine notions of pressure. This reengineering of frequency-based force is now coming full circle and completing the loop as the sensorium becomes the new territory that military and entertainment industries alike strive to colonise.

OSCILLATING PERCEPTION

Humans understand themselves, their orientation and their interactions within the world, as well as the world itself, via perceptions garnered from a number of physiological sensory organs and systems. Perceiving, transmitting and processing information available from the matrix of phenomena and stimuli within which humans are daily enveloped, the senses most historically researched, documented and tested are those of sight, sound, touch, taste and smell. This traditional cataloguing of the sensorial array can be traced back to Aristotle's *On the Soul* and its meditation upon the nature of living things and theoretical treatment of the five senses (Rorty and Nussbaum 1992). But while this assembly of perceptual systems no doubt works together to furnish the information we require to formulate a sense of agency, it is the information provided by our sense of sight that is privileged and dominates the construction of our understanding of the world. As Jonathan Sterne writes, 'Sight is in some ways the privileged sense in European philosophical discourse since the Enlightenment' (Sterne 2002, 3).

Historically underrecognised, at times marginalised and often misunderstood, the other perceptual systems—including the more recently scrutinised and classified senses of equilibrioception (balance), proprioception and kinaesthesia (joint motion and acceleration), time, nociception (pain), magnetoception (direction) and thermoception (temperature difference)—have not been chronicled in the same way as sight. Nor have they received the attention they deserve when one considers their importance in forming our conception of being in the world. As a result of our cultural adherence to explaining the known world via vision, the Occident has tended to regard the cartography, history and sociology of sonic percipience, experience and interpretation as a mere annex to the visual, as noted by German sociologist Georg Simmel in the early part of the twentieth century (Frisby and Featherstone 1997, 109–20).

Currently, discourses that define waveform spatiality and compose acoustic psychogeographies[4] are still quietly eclipsed from our collective lexicon and individuated praxis. *Sound Pressure* seeks to engage in and further the attempt to redress a sensorial balance within the philosophy of Western culture and beyond, to question, assay and challenge the disproportionate import afforded to the rationale of the ocular in favour of establishing the senses of hearing and touch as critical phenomenological indicators of urban spatiality, cultural chronology, collective psychological orientation, social relations and, ultimately, pressure. In doing so, the book seeks to establish a new form of sensorial agency that will simultaneously transmit and receive not only that which is perceptible but also that which is in constant oscillatory motion and often subsists on the verge of, or beyond, detection—the liminal, peripheral and viral.

OPERATIONS OF THE ANTENNA-BODY

The somatic model theorised throughout the book, in relation to the pressure that frequencies bring to bear upon it, is that of *the body as antenna*. This body is imbued with an agency that radiates from other contemporary mediated bodies such as those theorised by McKenzie Wark, bodies that, living in virtual geographies, 'no longer have roots' but instead 'have aerials' (Wark 1994, xiv). The notion of subjectivity that arises from the body as antenna simultaneously articulates the embodiment of two states of being—reception and transmission—but in doing so also begins to speak of an inherent *third* type of agency—that of being both, neither and more than these two capacities that partly compose its identity. It contends that, rather than being statically positioned, we exist and act between Manichaean poles of good and evil, sound and silence, place and space—constantly between stations and always changing channels.

Michel Foucault's conception of a subjectivity diffused in a web of power relations in flux—what might be called an oscillating subject[5]—is relevant to this premise, as are his propositions on the transformation of these power dynamics over time[6]. Such a vacillating subjectivity transmits, mutates and receives information according to the micropolitics, noise and collective harmonies emitted from the network of embodied speakers by which it finds itself surrounded. It is in this distributed system of social influence that we are able to monitor the ongoing spatial negotiations, methods of psychological alienation and strategies of physiological manipulation that simultaneously locate and displace our frequency-based agency. The transformative nature of the antenna-body that offers us new ways of perceiving sociopolitical relations in waveformed space.

FRANKENSTEIN IN FREQUENCIES

The antenna-body will therefore be a protagonist throughout this book, a lightning rod that allows us to speculate upon and synthesise past, current and future sound-system narratives. Echoing the temporal spectrum to be investigated, it is a waveformed mutant of sorts—a Frankenstein's monster in/ of frequencies—constructed from heterogeneous partial bodies of oscillatory thought. *Sound Pressure* demands that the hybrid body spatialised by waveforms[7] be grafted onto the wider body of Western philosophy (just as it was recognised that the 'body of space' needed to be inserted into the tradition of Occidental thought at the beginning of the last century). The mutant form we are going to construct here will be the nemesis of that postmodern pinup, the 'fragmented body', since it will transform a plethora of acoustic whispers into a coherent voice. In doing so, it will take part in a waveformed ecology that will be neither anxious nor precious about mixing perceptions, pressures and arguments from all fields of research—from neurology to geography, from sociology to legal studies and from musicology to thermodynamics. This conceptual operation will be performed, then, to better understand the origins of the antenna-body and to reveal its synthesis and its future modes of materiality.

The first contributor to this unsound surgery is a thinker who not only connects the spatialised body with the synthesised body but has also been instrumental in introducing his presence into the continuum of critical thought. It is from Henri Lefebvre that the first body part is transplanted in the form of his edict that space 'does not consist in the projection of an intellectual representation, does not arise from the visible-readable realm, but it is first of all *heard* (listened to) and enacted (through physical gestures and movements)' (Lefebvre 1991, 200). For Lefebvre, then, in the first instance space is heard rather than seen, and it is this essential assertion that will provide our synthetic body with its effectiveness as a cartographer and ecologist of waveformed environments. In a spatiality in which listening provides the predominant mode of cognitive association, it is Lefebvre's ear that will be grafted onto our waveformed holobody.

The next procedure concerns the replacement of the eyes, and it is to Michel de Certeau that we turn here, for the ocular system of the cornea, iris and pupil will not be the dominant form of sensory apparatus. According to his acute critique of our sensory comportments, the hierarchy of the perceptual apparatus requires a radical overhaul, for 'our society is characterized by a cancerous growth of vision, measuring everything by its ability to show or be shown and transmuting communication into a visual journey. It is a sort of epic of the eye and the impulse to read' (De Certeau 1984, xxi). De Certeau is explicit about the need for an alternative set of relations to be composed, through which to comprehend and speculate on our surroundings and the way we move through them. Since the mutated subject being built anticipates new modes of perception within an unfolding set of spatial relations, with the addition of de Certeau's critical eye, the antenna-body offers a communicable and affective channel through which to reimagine the practices of everyday living[8].

So that our body is fully aware of its potential to be transformed anew, its expansive cognitive faculties will be supplied by British anthropologist Gregory Bateson, who insisted that the 'mental world—the mind—the world of information processing—is not limited by the skin' (Bateson 1973, 429). For Bateson, perception and activity are not orchestrated solely by the brain and merely delegated to the fleshy machinery of the body. They are contingent on the fluctuating presences of the subject within an environment and the relations orchestrated by such a form of embodiment. Our synthesised antenna-body will never be limited by its dermic interface.

Its intramural dimensions will therefore resonate sympathetically with the overall constitution of the body that, supplied by Michel Serres' parasite, will be essential for the subject's transmissible disposition, since all the newly grafted parts are dependent on their relations with the new host body as well as with each other. Thus the flows of the body will be viral, mixing together the new surgically sutured waveformed parts; any notion of a prosthetic anatomy will be negated as the antenna-body's viral plasma evolves it into a coherent form, without borders and inviting further augmentation. This is not a stitched-together body so much as a positively infectious body—a subjectivity full of viral association, mutation and thirded extremities.

In this way, the antenna-body is analogous to Serres' (1982) 'pest': both embody the capacity to resist, divert and assimilate the subjugating technologies and political pressures that writers such as Foucault identify as controlling the somatic. Serres' notion of parasitic noise as a third entity within all communicative acts is a crucial proposition in terms of the discourse and potential convergences it initiates between waveformed and viral theory. The formatted relationships between music and the mass social body discussed in the first two chapters adhere to Serres' description of noise as an inevitable, relationally composed thirded presence, but in chapters 4 and 5 HyperSonic Sound System technology comes to refute such suppositions, and noise/the 'third man' is theorised as being a third voice that is *channelled into* the subject. Thus, rather than being exteriorised in acts of communication, the noisy, thirded viral subjectivity is theorised as constituting a psychological munition whose payload is transmitted into the very inner sanctum of communication: the subject's skull.

Our protagonist's essential instincts will be replaced by Baruch Spinoza's appetites: 'a body's conatus, or striving to persist in its power to affect and be affected, its potential. Whereas instinct usually denotes a closed, preprogrammed system with no room for change, appetite is future-facing and always in conjunction with the body's relation to a shifting ecology, its open-ended relationality' (Goodman 2009, 70). If our mutating subject is to transmit back to us information about the diverse range of modulating factors within the new waveformed ecology, it will require such a system of innate sensibilities. For we need to provide our body with an unflinching ability to connect and mutate in relation to its environment and to the pressures exerted upon it so that 'the focus shifts from what a body is, even in its technologically extended sense, to its powers—what it can do' (36). And in order for our antenna-body to be able to question *where it can be*, it needs a curiosity—an intense desire to understand the hinterlands of the soundscape[9] and all those phenomena that vacillate at the edges of perception. To donate to us our final faculty, that of curiosity, we turn to radio-wave researcher Heinrich Hertz[10], who 'spoke of the "narrow borderland of the senses" between consciousness and the "world of actual things". For it was Hertz who declared that for a "proper understanding of ourselves and of the world it is of the highest importance that this borderland should be thoroughly explored"' (Johnson and Cloonan 2009, 13).

As stated, the antenna-body will need transforming further so that it might amplify, record and modulate the waveformed ecology. The operations undertaken will not, however, deliver the somatic from its fleshy and intense relations with the world in which it finds itself. For what we are proposing here is a frequency fiction of the carnal, one that steers clear of the fetishistic transhumanist[11] predictions of writers such as Hans Moravec (2000) who eulogise an eschatology of escape from the body (and its difficulties) by way of a scientific and technological upgrade of the human. We are instead composing a body that listens for the ways in which political technologies of the somatic are implemented, as it is these augmentations and strictures that threaten to 'invest it, mark it, train it, torture it, force it to carry out tasks, to perform ceremonies, to emit signs' (Foucault 1975, 25). When the protagonist of our frequency fiction can detect such perceptible and imperceptible external pressures, when it can hear the unhearable, touch the untouchable and when its viral engineering can predict future forms of sociospatialised organisation, then it will have heeded the call of the waveformed wild—a clarion call to arms, but also to ears, skin, hair, bone, neurons, muscle and nerves, a call to mutate: 'an imperative to grow new organs, to expand our sensorium and our body to some new, as yet unimaginable, perhaps ultimately impossible, dimensions' (Jameson 1991, 80).

PROPOSITIONS

A mutation of a different sort takes place between chapters 3 and 4: a break in the episteme of speaker-system ideology that occurs between the deployment of sonic torture in Guantánamo Bay and the emergence of ultrasonic agency. It is posited that the torture techniques deployed in Guantánamo Bay are the culmination of repetitive strategies that, ever since it became possible to record and play back sound via the phonograph in the 1870s, have become significant waveformed tropes within Western culture. The repetitive radio playback of songs on heavy rotation[12], a sales technique deployed by record companies, has been co-opted by military organisations and taken to its logical extreme: increasing rotation while cutting all silence or 'chat' between musical transmissions transforms an infectious sales technique into a torture weapon for contaminating the rational mind.

Considering repetition to be the most conspicuous organising principle of twentieth-century production, storage, distribution and social control, French economist Jacques Attali writes that 'the emplacement of general replication transforms the conditions of political control. It is no longer a question of making people believe, as it was in representation. Rather, it is a question of Silencing—through direct, channelled control, through imposed silence instead of persuasion' (Attali 1985, 121). The repetitive playback of Western music genres such as heavy metal, disco and country over a period of many days and nights was designed to deprive Guantánamo Bay's detainees of sleep, to disorient them and ultimately to make them lose control of their minds. Embodied in torture practices, this process darkly reflects and distorts the sentiments expressed by Attali three decades earlier.

Such assimilation by the military of techniques originating in the civilian domain and vice versa are analysed throughout the book in order to understand how waveformed strategies and techniques are programmed and transmitted between these host bodies. The co-opting of popular music as a torture weapon is emblematic of a more generalised absorption of cultural practices, products and tactics into the modus operandi of military organisations. The 'installation-art' techniques utilised in military training camps in the United States and Canada (Heys and Hennlich 2010) and the harnessing of the philosophies of writers such as Paul Virilio and Gilles Deleuze and Félix Guattari by the Israeli Defense Forces (Weizman 2006) are representative of an organised mobilisation of cultural ideology into martial practice. It is hypothesised that this radical shift—implemented by the military-entertainment complex in the way that culture functions (including products, ideas and behaviours)—means that the production of culture can no longer be assumed to be inherently aligned with the resistance, antihegemonic and left-wing ideology with which it has traditionally been associated. Put simply, it is contended that, post–Guantánamo Bay, it is difficult to think about music (and by extension culture) in the same way again.

Another of the book's significant arguments proposes that, from the 1920s on, the military-industrial and military-entertainment complexes have succeeded in embedding within capitalism a network of waveformed techniques and strategies (themselves indicative of wider associated cultural practices) that serve to separate, isolate and alienate the subject from the social body of which it is a part. From the industrial factory and its workforce (chapter 1) to the religious compound and the extended family of the sect (chapter 2), waveforms have been deployed to manipulate the behaviours of ever-smaller numbers of subjects on ever-decreasing spatial scales. The conclusion of this strategy is reached with the example of the torture cell and the isolated detainee (chapter 3).

Subsequently, unable to reduce any further the space and number of the targeted subjects, new waveformed techniques and technologies—emblematised by the HyperSonic Sound System (chapter 4) and the flat-panel speaker (also referred to as the Whispering Window, chapter 5)—move beyond surround-sound environs into the inner-cranial spatiality of the subject's skull. While the individual subject remains a target of martial waveformed instrumentality, the external spatiality within which she was previously affected is now inverted. The ultrasonic beam and bone-conduction techniques foreshadow the new, alienating environment that its target will inhabit: an internalised future geography of neural flows and signal pulses.

In virtue of these arguments, what *Sound Pressure* calls for and instigates is a practice of frequency-based cartography—a mapping of the soundscape's territorialisation by the military-entertainment complex. Such a discipline will resemble a waveformed psychogeography, heuristically charting the spatial concerns of the neural realm as well as the organic, material and built environment. As a field of research it will have a wide-ranging remit to explore the spatial, psychological, physiological, social, economic and sexual effects that frequencies have upon our subjectivity and agency. Its methodology—as suggested through the structuring of this book—is multidisciplinary and multichannel, creating new forms of knowledge about phenomena such as LRADs[13], Mosquitos[14], intonarumori[15], loudhailers and Sequential Arc Discharge Acoustic Generators[16]—all components of the metanetwork of speaker systems through which the rhythms and cadences of power are transmitted, connected and modulated.

SONIC GEOGRAPHY

Violence, force and pressure expressed through waveforms extend the interests and agency of those who transmit them. This implies that frequency-based pressure is inextricably linked to the notion of extension and, by association, geography; for waveformed pressure is always pushing into, on and through space and that which inhabits it. If we speak of waveformed pressure then we inevitably suggest a dynamics of competition over space, whether psychological, geographical, social, sexual or economic in nature—or more likely some combination of all of these. In view of the recently purported 'acoustic attacks' on staff of the American embassy in Havana (2017) and Guangzhou (2018), this is an opportune moment to further investigate the associations between waveformed pressure and spatial theory.

This approach is indebted to the investigative modalities proposed by geographers such as Lefebvre in texts such as *The Production of Space* (1991; originally published in 1974) and by Foucault in his essay 'Of Other Spaces: Utopias and Heterotopias' (2002; originally published in 1967). Along with other thinkers such as David Harvey (2001), Lefebvre offers philosophical whisperings of how a methodology for a waveformed psychogeography might sound. In articulating his theory of *rhythmanalysis*[17], he proposed that space is perceived, in the first instance, by the ear (Lefebvre 1991, 200). Since waveforms have to move through space in order to be perceived, more than any other phenomena they best articulate the essential transitory nature of spatiality and the inherent violence of transgressing boundaries.

Edward Soja's 1996 text *Thirdspace: Journeys to Los Angeles and Other Real-and-Imagined Places*[18] is also central to the way in which the spatial approach of *Sound Pressure* has been composed. Soja devises a trialectical[19] approach based on Lefebvre's (1991) notion of the spatial triad in order to insert spatiality back into the lexicon of Western philosophical thought. With his methodological stance of 'thirding as othering'[20], Soja proposes to dislocate dualistic modalities of thinking that had previously consigned spatial theory to occupying a mere footnote to time-based readings of the world and of human presence and movement within it. Here his ideas are extended into the realm of sonic theory in order to complexify our comprehension of waveforms, which are too often posited in terms of binary relations such as noise/music or pleasure/dissonance.

This implementation of Soja's methodology of *thirding* is crucial to the arrangement of the five case studies explored here. Soja's technique is to employ a discursive geographic instrumentality as an analytical tool to unearth newly occurring relations between periphery and centre. Since this text explores both those waveformed phenomena that are central to our comprehension of the world—the sonic—and those that exist at the edges of our perception—the ultrasonic and infrasonic—it is essential that our mapping also extend to encompass the hinterlands of waveformed phenomena. Drawing on theorists such as bell hooks, Soja recognises that the periphery is where modes of resistance manifest themselves. His exploration of phenomena and activities that oscillate on the fringes is important in investigating how the boundaries of social agency vacillate and can be subverted or governed by waveformed techniques.

Throughout the text, a number of approaches are employed to determine the key dynamics involved in the application of pressure in space. As a compressed example of the multiple discourses involved in this enterprise, it is useful here to offer an everyday example of how waveforms compose realistic mapping systems. It is a given that, 'on the basis of pressure variations at your ears, you gain access to an abundantly detailed world of sounds, things, and happenings' (O'Callaghan 2009a, 580). A schematic portrait of our exemplary subject would then simply list the various origins of these variations in pressure as she sits in front of her computer with music playing. As she writes, she can also hear the low whirring of the hard-drive fan, traffic noise, people shouting and birds calling from the environment outside the window of her apartment, the humming of the fridge, the internal oscillations of her cilia courtesy of tinnitus and the faint drone of jet engines eight miles above.

On a daily basis, this subject is simultaneously a part of six, seven or eight sonic spatialities at any given moment—and we have not even begun to take into account the influences of ultrasonic and infrasonic waveforms upon her moods, rhythms and perceptions. Sonically speaking, she lives (to paraphrase de Certeau 1984, 201) in a pile of heterogeneous spaces, each with its own particular dimensions of social presence, mobility and exclusions. Within this dense mix of waveformed spatialities, 'the central problem of auditory perception involves the auditory system's capacity to discern from complex wave information the number, qualities, location, and duration of sounds and sources in one's environment' (O'Callaghan 2009a, 580).

Significantly, a conception of movement through the frequency-based domain as being choreographed simultaneously by presence, mutated transmission, slippage and convergence tends to echo the language and ideas employed to describe our existence between the colliding worlds of communications technologies. We may hope for clear and direct lines when utilising mobile phones, videoconferencing software and satellite uplinks, but in actuality our remote exchanges are supplemented by other realities: noise, interference, signal drop and crossed connections. In contrast to the more basic perceptions of the ocularised environment, the sedimentary nature of waveformed-communications space seems to reiterate and compound these concerns pertaining to existence in a multitiered space: a proposition supported by Julian Henriques' assertion that 'sonic time, like sonic space, is not travelled in straight lines. It's too heterogeneous for that. If written it would be in layers and with the depths of a palimpsest. In contrast, the writing of visual space is in Euclidean straight lines and flat planes on the uniform blank expanse of a *tabula rasa*' (Henriques 2003, 459).

Henriques' textual invocation of the sonic strata not only explicates the deep lattice of waveforms that constitute the sonic domain, it also delivers us back, for a final time, to the spatial languages proposed by Lefebvre. Through his concept of 'meshwork', Lefebvre (1991, 117–18) describes how the cadence of movements within the lived environment belies the textural dynamics of spatiality. Applied to the soundscape, the concept of waveformed texturality associates the ways in which we slip in and out of frequency-based environments with the manner in which we effortlessly transverse the remote spaces opened up by communications technologies. This helps explain why both academic and nonacademic interest and research into the dynamics of the sonic have dramatically intensified over the past decade: theories of waveformed spatiality offering the most coherent translations of our simultaneous presence (and the self's nonpresence) within multiple spaces and of the processes of transformation and divergent temporalities associated with these spaces.

As technology allows us to be increasingly mobile and to disperse ourselves into lateral mobile networks, distant topographies and far-flung relationships, our comprehension of twenty-first-century existence equates to being continually enveloped in multiple realities (this proposed dispersal and nonpresence of the self echoing Friedrich Kittler's [1999b] notions about the manner in which the modern self has been composed through mediating technologies). We are always only a mouse click away from the digital space of the Internet, a button press away from the remote verbal space of the cell phone and a randomising shuffle away from the emotionally modulating space of the MP3 player. Accordingly, this book presents a sequencing of events that informs us about the ways in which frequency-based technologies and communications networks have come to symbiotically negotiate a spatial fascination that has always been with us—a compulsive problematic that equates our capacity to be anywhere at any time with our potential to become sonically omnipresent.

Indeed, the capacity to have one's voice or sounds present in a diverse range of locations at the same time resonates throughout this research. It is inferred that the theoretical dominance, throughout Western philosophy, of temporality over spatiality is challenged by waveformed ontologies. 'We have often been told . . . that we now inhabit the synchronic rather than the diachronic,' states Fredric Jameson. He argues, however, that 'it is at least empirically arguable that our daily life, our psychic experience, our cultural languages, are today dominated by categories of space rather than by categories of time' (Jameson 1995, 64). The lived experience of sonic spatiality proposed throughout the text questions our assumptions not only about the presupposed linearity of space and time but also about the points at which they intersect and pull each other out of rhythm and shape.

As has been established, this text argues from and for a frequency-based perspective on the five speaker systems that constitute the focus of the case studies contained herein: from Muzak's wired-radio speaker system to the FBI's surround-sound system at Waco, and from the darkly intimate speakers used in Guantánamo Bay to the covert technologies that constitute the HSS and the Whispering Window. In terms of the import of spatial relations to the functionality of these sound systems, Edward Said's point in *Culture and Imperialism* (1993), that we are never entirely outside of the concerns of geography, is well made. All of the frequency-based technologies considered here actively renegotiate the limits of inclusion, exclusion and transgression for those within earshot. The topology[21] developed from their analyses augments the channels that lead from the outside in as much as those that lead from the inside out.

SOUND AS VIRUS

Nothing encapsulates the anxiety of the liminal (the undetected) and its potential to transgress the periphery and travel into the core—of networks, speaker systems, bodies and language—quite like the virus. Numerous terms are used to describe the multitudinous carriers of viral culture: *worms*, *malwares* and *Trojans* within information systems; *bacteriophages*, *microorganisms* and *pathogens* within the corporeal matrix of fluid and airborne exchange; and *earworms*, *influencers* and *sneezers* within the social networks targeted by viral-marketing strategists who search for individuals with the most *social-networking potential* (SNP) so that they can be persuaded to transmit the popularity of demographically relevant products[22]. And the oscillating thresholds of unsound should also be included in this taxonomy of the liminal, the barely sensed—especially when deployed on unsuspecting targets.

The notion of the barely sensed also alludes to the thresholds where the living, the living dead and the dead intersect and to the conviction, expressed in many cultural products, that the virus and the waveform can open up conduits between these states of being. Viruses, like waveforms, are difficult to control, map and direct. Both have the capacity to move imperceptibly, to enter without permission, to create networks of affect, bodies of evidence and psychological states of being. The virus furtively penetrates computer systems, renegotiates and rewrites network protocol on the fly and creates cultures of fear and paranoia simply at the mention of its name. As for waveforms, we might refer to the long-range waveforms of infrasound and their propensity to travel thousands of miles unnoticed by humans, the capacity of ultrasound to scan and render visible that which cannot be perceived by the naked eye, and the anxious culture of fear and suspicion that envelops the waveformed body and its vulnerability to the power of frequencies. Both waveforms and viruses have the capacity to propagate behaviours of a transgressive nature. The fact that these behaviours are orchestrated within networks where it is difficult to perceive at what point the zones of the living and the dead begin and end renders them potentially hazardous.

Eugene Thacker (2004) analyses the mediated links between living and dead networks, observing that 'the horror of contemporary "living dead" is not just the fear of being reduced to nothing but body, but, in the "network society", perhaps the horror of the "living dead" is the fear being reduced to nothing but information—*or not being able to distinguish between contagion and transmission*. In this sense the paradox of the living dead is also the paradox of "vital statistics", a sort of living dead network that exceeds and even supercedes the "bare life" of the organism' (emphasis mine). If we take this meditation upon the disappearance of the body into data and the associated fear of losing the capacity to perceive the somatic threshold of presence and transfer it to ultrasonic, infrasonic and sonic networks, it provides further useful pointers as to the liminal nature of frequencies and their connection to the viral. We predominantly perceive hauntings in terms of the ability of a vibratory entity to discreetly penetrate the network of living things through manipulation of voice and object. Our fear is fed by the threat of contact and infection from liminal presences, of our 'selves' being transmuted into waveforms. It is the difficulty we encounter when trying to contain, control and map the ephemeral nature of the viral and the waveform that renders them such anxiety-provoking phenomena, reminding us as it does of our existence in the oscillating thresholds between the sonic and the silent, the transitory and the static, the living and the dead.

In more speculative terms, the viral can be thought of as a mode of communication, a discourse that mutates and propagates meaning within and between technological, somatic and social networks, the enfant terrible of a capitalist system that reproduces itself via channels of total transience, sublimely articulating the breathless motion and inevitable crisis of stasis within systems of exchange. It is at once the perfect model of a socio-economic system conducting its business through the flexible organs of distributed agency and the potential saboteur of that very same system. Analogously, music is potentially both 'the medicine of the mind' (Logan as quoted in Storr 1992) and its poison.

WAVEFORMED LANGUAGE AND LISTENING

Finally, how do we make things communicable—so that they can be uttered in the first instance and subsequently modified and heard? To begin with, a perceptual consideration: We cannot hear the act of hearing. We can watch another's act of looking and form an opinion about the ways in which they direct their gaze, but we cannot listen to another's act of listening. The mode of listening is a mute interaction. Hearing is therefore a sense that offers refuge to the covert and the liminal, creating a contradictory sensory spatial dynamic that is hidden and unutterable but whose boundaries are porous and malleable. Accordingly, when describing what is heard, we have to admit other types of discourse into the analytical space occupied by physics, phenomenology and neurology. The territorial borders of the perception of waveformed space cannot therefore be effectively shut and sealed by facts, observation or recording.

The fact that the unutterable and the unspeakable are intrinsic to such environments and must constantly be negotiated by listeners means that the boundaries are constantly being recomposed and are never rigid. And it is precisely this fluctuating dynamic that permits unofficial, unscientific and unrecognised languages to translate waveforms into exchangeable knowledge—narratives that shred and dislocate the definitions of those translations at the same time. In such circumstances as the alleged 2017 acoustic attacks on the American embassy in Havana, the oscillatory and transient nature of 'meaning' becomes apparent. Conspiracy theories, memes, quasipolitical conjecture and science fantasy all come into play to explain what happened to the staff members who suffered ailments ranging from nausea to mild cerebral trauma. These marginal discourses reside alongside 'official' bodies of knowledge, rubbing up against them, irritating their rhythmical methods of erudition, their forms of rationalisation and their epistemological suppositions.

Throughout the following five chapters it is contended that we have not yet fully developed a wider socially accepted language to explain the topological characteristics, psychological orientation and physiological violence of waveforms. And where there are no available words to explain phenomena, whether frequency-based, material or action-based, this leads to such phenomena being freighted with meanings that belong to the anxiety-ridden registers of the otherworldly and the hallucinatory. These displaced and misunderstood interpretations are often not translatable into the existing taxonomies that are supposed to evaluate and assess them. As a result, narrative and interpretative discourses explaining frequency-based phenomena such as the Cuban affair become marginalised and end up being annexed to cultural expressions of anxiety and paranoia.

Deprived of the ability to label and designate phenomena, we find ourselves momentarily dislocated, unable to situate ourselves as rational, centred subjects. Our positions, as subjects with an aptitude to sense, judge and implement our ideas upon our environment, based on the understanding that we are able to know and to name, are threatened when we are unable to articulate that which we perceive. It is the unspeakable that we consign to our subconscious and actively try to forget, for that which we cannot remember to speak of is that which is half dead to us. It is here between the transfer of the signal and the monotone of the flatline that waveformed memories operate, as conduits between the articulated and the unutterable, in the living-dead networks of perception.

There is, however, a power in being located on the periphery of the living and the dead. The power of frequencies partly resides in their disorienting capacity to displace the language, description and perception of both states of being. In spatial terms, Foucault names the cemetery as the place 'connected with all the sites of the city, state or society or village' (Foucault 2002, 233). It is in this transitional locale that he recognises a shift occurring, a transmutation of the understanding and cartography of the living and the deceased. Because, 'from the moment when people are no longer sure that they have a soul or that the body will regain life, it is perhaps necessary to give much more attention to the dead body, which is ultimately the only trace of our existence in the world and in language' (ibid.). It is this same sense of uncertainty that prompts us to characterise waveformed space as a refuge of the tenebrous, an ambiguous zone harbouring phenomena and interactions we are unable to rationalise, whether it be sounds at night that cannot be explained—interpreted as movements of the dead—or the 'inner voices' that we ascribe to the mentally ill to help us understand the noisy multichannel nature of schizophrenia[23].

Taking this tendency—to ascribe the sonic to the otherworldly—to its nefarious conclusion, we can track down the example par excellence that bespeaks our cultural anxiety and disquiet concerning the purgatorial power of frequencies. Allegations (often made by organisations affiliated with Christianity) of backward-recorded messages on vinyl records—also known as *backmasking*[24]—reveal the full extent of our moral, social and bestial fears about music's capacity to channel information from perdition, fears that are only fully sated when *subliminal messages*[25] are afforded the capacity to propel mentally unstable subjects such as American serial killer Richard Ramirez to perform gruesome acts of violence. This anxious disposition subsequently attributes music—and by extension frequencies—with the potential to manufacture evil deeds and, more than that, with the power to transfer the somatic and the spiritual into the environs of the underworld itself. In this context, music can be perceived as a phenomenon that operates in the passage between psychological torment and its physical expression, between the scientifically monitored condition and the unthinkable act, as a force that transgresses the material world of things yet deeply affects actions within it. Thus it is music's contradictory symbolic index—as religious celebratory expression and as transmission of the Devil's will—that renders waveforms as phenomena to be both feared and revered.

Since the emergence of the first wired speaker system in 1922, the role of such systems in shaping emotional, sociopolitical, aesthetic and violent registers has become increasingly abstruse and increasingly compounded with the ontology of the as-yet-unperceived. *Sound Pressure* has been written in a manner that echoes the ways in which presence, territorialisation and affect within the waveformed domain have mutated and shifted, moving from the easily perceptible and graspable history of Muzak to the silent conspiracy-bound futurism of the directional ultrasonic beam and the Whispering Window. As technological enhancements such as the HSS have rendered the twenty-first-century soundscape an increasingly abstract space for the sensorium to inhabit, so the language and ideas chosen to document this trajectory become increasingly speculative over the course of the book. Thus, the level of abstraction within the text echoes the abstraction of speaker-system usage within the frequency-based domain—from historical documentation of industrial forms of sonic organisation to futuristic speculation upon the ways in which the sensory apparatus will become mutated, exteriorised and territorialised by unsound techniques.

NOTES

1. The first electrically powered speaker-system network transmitting radio programming to the workers of industrial factories in the early part of the twentieth century.

2. The use of the terms *strategy* and *tactic* throughout the text is indebted to Michel de Certeau's reinterpretation of them outside of their military meanings, as implicated social techniques. He defines strategy as a type of embedded behaviour that is enacted by an empowered authoritarian source such as an organisation, corporation or government, while a tactic is a type of action carried out by a more nomadic and less formally organised social group or individual, usually out of necessity, but without any intention to 'take over' or 'sabotage', which differentiates it from modes of resistance such as guerrilla warfare (De Certeau 1984, 29–42).

3. The notion of unsound as a vibrational cosmology of affect and preemptive embodiment at the margins of perception is further discussed by Steve Goodman throughout his book *Sonic Warfare: Sound, Affect, and the Ecology of Fear* (2009).

4. A phrase initially coined by Guy Debord in his 1955 short text 'Introduction to a Critique of Urban Geography'. In his words, psychogeography constitutes 'the study of the precise laws and specific effects of the geographical environment, consciously organised or not, on the emotions and behavior of individuals' (Debord 1955).

5. Proposing that the individual does not merely act as a passive receiver but also as an active proponent—or transmitter—in a network of power relations, Foucault states that 'power is employed and exercised through a net-like organisation. And not only do individuals circulate between its threads; they are always in the position of simultaneously undergoing and exercising this power. They are not only its inert or consenting target; they are always also the elements of its articulation' (Foucault 1980, 98). In this way, the receiving and transmitting capacity of the oscillating subject constitutes the self as what we are calling an *antenna-body*.

6. Foucault's mode of analysis was dependent upon ascertaining how changes and differences occur over time, with transformations occurring within a context that allows some mutations to pass through while impeding others. Thus Foucault's methodological approach is a historical one, albeit that of a radicalised history, that uncovers the way in which truth is conceived and how circumstances are engineered to authenticate and propagate certain discourses, techniques and, ultimately, truths.

7. By *waveforms* we understand all frequencies within the infrasonic, sonic and ultrasonic ranges.

8. A reference to Michel de Certeau's proposal that there is a potential subversive agency within the everyday practices and articulations of the mundane and the ordinary—what he called the 'tactics of living' (De Certeau 1984).

9. Coined by R. Murray Schafer in *The Tuning of the World* (1977), *soundscape* refers to the enveloping world of acoustic phenomena that includes each and every environment within which we might find ourselves (except a vacuum).

10. A German physicist who pioneered the testing and recording of waveforms outside of the audible range of human hearing, Hertz constructed a device to both compose and detect VHF and UHF waves and proved to the world at large that they existed. *Hertz* subsequently became the unit of measurement of frequency (in cycles per second).

11. Also known as *posthumanism*, this intellectual and cultural movement born in the 1980s supports the enhancement of the human form by biotechnologists (among other future engineers of the body). Proponents of transhumanism believe that somatic phenomena such as disability, aging and death are avoidable, leading many critics to liken its eugenic values to the ideas of proponents of 'master race' ideologies.

12. The repetitive playing of songs on the radio over a period of weeks or months in order to maximise their sales potential.

13. Developed by the LRAD Company in San Diego, the Long Range Acoustic Device speaker system has the capacity to transmit sound over three thousand metres, with a maximum range of 1250 metres over eighty-eight decibels of background noise. On the LRAD website in 2012, the producers of the technology declared that 'it is used to prepare the battlefield, or prepare city spaces or possible sites of conflict in Iraqi cities for military operations such as Fallujah and Baghdad, almost like an acoustic cultural swab. It was . . . used against protestors in 2004 in Union Square, New York'.

14. An ultrasonic speaker system that emits frequencies only perceivable by teenagers, the Mosquito is utilised by shopkeepers, petrol-station companies, railways and governments to deter young people from 'loitering' on their premises or within public areas.

15. *Intonarumori* (literally 'noise intoners') were instruments created by Futurist musical composer Luigi Russolo in 1913 designed to orchestrate a wide range of sonic textures comparable to those produced by the machines that the Futurists so revered.

16. Also known as SADAG, a sonic weapon that produces high-intensity impulsive sound waves by purely electrical means.

17. Many theorists have deployed Lefevbre's concept of rhythmanalysis in an effort to define the mutating experiential relations created as the (not-always-present) body moves through space. According to Crang and Thrift (2000, 22), with this term Lefebvre 'sought to convey a quality of experience which arises out of the conglomeration of different spaces and times, sometimes in harmony, sometimes in discord, but always mobile—encountering—alive, to be found in modern societies. . . . He perhaps came closest to producing a sense of an embodied, inhuman, travelling means of inscription'.

18. Edward Soja conceptualises *thirdspace* as a tenet of his postmodern geography. The term refers to spaces that are at the same time both imagined and real: 'Thirdspace is a purposefully tentative and flexible term that attempts to capture what is actually a constantly shifting and changing milieu of ideas, events, appearances and meanings. . . . There is a growing awareness of the simultaneity and interwoven complexity of the social, the historical and the spatial, their inseparability and interdependence. . . . The challenge being raised in *Thirdspace* is therefore transdisciplinary in scope. It cuts across all modes of thought' (Soja 1996, 2–3).

19. A model of trialectics that is based on Soja's reinterpretation of Lefebvre's concept of the spatial triad is employed throughout this book. In Soja's trialectic system, there are three forces influencing each other, negating the fixed nexus of sublation located by Hegel in his dialectical model (in a Hegelian framework, conflict between thesis and antithesis is resolved by the synthesis). In Soja's words, 'the starting-point for this strategic reopening and rethinking of new possibilities is the provocative shift back from epistemology to ontology and specifically to the ontological trialectic of Spatiality-Historicality-Sociality. This ontological rebalancing act induces a radical scepticism toward all established epistemologies. All traditional ways of confidently obtaining knowledge of the world' (Soja 1996, 81). In the current context, a trialectic model of analysis is used to initiate new approaches to interpreting waveformed spatiality and sociality.

20. The concept of *thirding* as a methodological and analytical tool used to dislocate dualistic thinking has been explored by a number of writers. Of particular note here is Michel Serres' proposition that noise is the inevitable 'third man' between two connecting parties. He proposes that within an exchange of sonic information, noise is located in an ambiguous position, being both a peripheral and central founding presence in the formatting of such communication. Accordingly, Serres writes that 'we are surrounded by noise. We are in the noises of the world, and we cannot close our door to their reception. In the beginning is noise' (Serres 1982, 126).

Also useful here is the thirding dynamic employed by Hélène Cixous in her novel *The Third Body* (1970). She introduces it (on the very first page) in order to define the somatic relationship between two lovers, rupturing the dualistic understanding of agency and sublimation as the two bodies, at once the same and other, ultimately construct a third identity. For Homi K. Bhabha (1994), thirdspace represents a hybrid spatiality of antagonism, constant tension and pregnant chaos. He argues that from this standpoint one can destabilise the binary oppositions that construct the First and Third Worlds, including those of centre/margin, civilised/savage and capital/labour. Through this theoretical act of associative dislocation, he suggests that by employing thirding techniques it would be possible to reconstruct discursive political discourses, which would help disempower systems of colonization.

21. Steven Connor's reading of Michel Serres' spatial analysis of topology is referenced and utilised throughout this book. For Connor, 'Topology may be defined as the study of the spatial properties of an object that remain invariant under homeomorphic deformation, which is to say, broadly, actions of stretching, squeezing, or folding. [It is] not concerned with exact measurement, which is the domain of geometry ... but rather with spatial relations, such as continuity, neighbourhood, insideness and outsideness, disjunction and connection. ... Because topology is concerned with what remains invariant as a result of transformation, it may be thought of as geometry plus time, geometry given body by motion' (Connor 2004, 106).

22. Also referred to as viral advertising, this marketing technique identifies influential subjects and social networks to transmit brand awareness through, by using propagative methods analogous to those employed by digitally and somatically based viruses. One of the first books to explore the media phenomenon of viral marketing and still one of the most salient is Douglas Rushkoff's *Media Virus! Hidden Agendas in Popular Culture* (1996).

23. A clinical understanding of schizophrenia is employed in these pages, as opposed to Deleuze and Guattari's notion of schizophrenia developed in texts such as of *Anti-Oedipus: Capitalism and Schizophrenia* (2004).

24. Numerous popular recording artists have been accused of utilising backmasking techniques on their records, including Britney Spears, ELO and Eminem. The most infamous incident of a defendant alleging that backward masking on a record had inspired their actions occurred during the 1970 trial of Charles Manson for the Tate/LaBianca murders committed the year prior. During judicial proceedings, the defence proposed that Manson believed an apocalyptic race war would engulf the country and that the Beatles, through songs such as 'Helter Skelter' (1968), had embedded hidden messages foretelling such violence. Manson's delusional response to these perceived messages was to record his own prophetic music (Manson 1970) and to murder Leno and Rosemary LaBianca and actress Sharon Tate (among others) in order to trigger the supposed conflict.

In 1985, psychological tests were conducted by Vokey and Read to ascertain whether subliminal messages in music affect behaviour in any way. Their results concluded that there was no evidence to support such an idea and that the perception of such messages in music tells us more about a subject's will to fabricate than it does about the actual existence of the implied content (Vokey and Read 1985).

25. Purported transmissions or communications embedded in musical recordings (and sometimes in other cultural products such as films and artworks) that covertly urge audience members to carry out (often violent) actions. Examples of subliminal messages being blamed for violence include the court case of US serial killer Richard Ramirez, convicted of thirteen murders by a Californian court in 1989. A fan of the hard-rock group AC/DC, Ramirez cited their *Highway to Hell* (1979) recording as his favourite album. David Oates, a reverse-speech advocate, maintained that subliminal messages on this record—such as 'My name is Lucifer' and 'She belongs in hell'—had driven Ramirez to commit the murders.

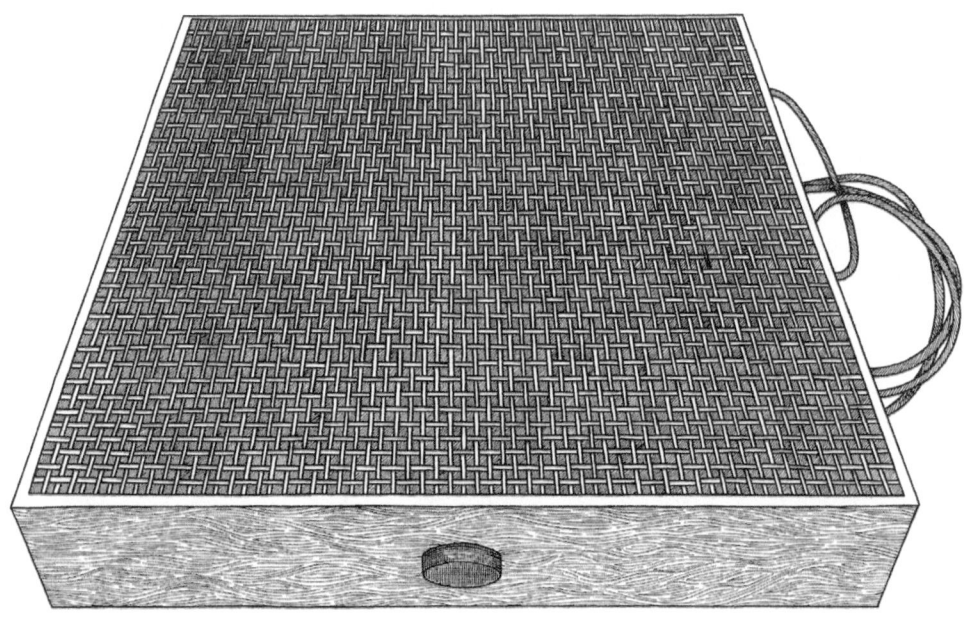

Jensen CN-82 2 Way High Fidelity Speaker. *Illustration by Krystian Griffiths*

Chapter One

Muzak's Influence in the Fordist Factory

DISTRIBUTED RHYTHMS OF LABOUR

Following the onset of the Second Industrial Revolution[1], the early part of the twentieth century saw a boom in the mechanisation of European factories owing largely to the demands exerted upon the rhythms of agricultural, economic and labour production by the advent of World War I in 1914. Manuel DeLanda alludes to the national requirement for technologically fluid manufacturing systems in such times of conflict, writing, 'As the last two great wars have shown, victory goes to the nation most capable of mobilizing its industrial might' (DeLanda 1991, 34). As war compounded industrialisation's drive to organise large masses of bodies into a workforce, the cultural and political will to maintain a steady production of goods and techniques to improve industrial efficiency became of paramount importance. The social sciences were harnessed in this effort to organise and systematise the most economic ways for the individual and mass social body to carry out tasks in the workplace. Methods by which to situate, order and discipline bodies en masse—what Michel Foucault called the *body politic*—had already been institutionalised within the prison system and would subsequently be transplanted into industrial environments. Foucault defines this somatic ordering as 'a set of material elements and techniques that serve as weapons, relays, communication routes and supports for the power and knowledge relations that invest human bodies and subjugate them by turning them into objects of knowledge' (Foucault 1975, 28). Nowhere is the desire to produce the industrialised body as a knowable and controllable asset more apparent than in the 1911 monograph *The Principles of Scientific Management* by American mechanical engineer Frederick Winslow Taylor (1911). It outlined a set of techniques that would become synonymous with Taylor, which were to be highly influential in shaping twentieth-century capitalism and continue to influence workplace procedure today.

Taylorism, influenced by military command structures and by the inventions and writings of Charles Babbage (see Babbage 1835)—especially his championing of the benefits of the division of labour (Jamieson 1991, 60)—was dedicated to the organisation of bodies and the maximisation of their labour potential. As DeLanda states, 'The methods the military developed to shorten the chain of command were ... exported, through people like Frederick Taylor, to the civilian sector' (DeLanda 1991, 229). In the United States, industrialist Henry Ford was among the first to employ techniques determined by Taylor's time and motion studies[2]. When Ford discovered Taylor's research, he had already initiated and developed the installation of moving assembly belts in his automobile factories, yielding a substantial escalation in production and sales.

While it is accepted that the succession of transformative organising principles implemented by Ford became the dominant model for the practice and comprehension of mass production and consumption in US capitalism between the 1940s and the 1960s, it could be said that, for the workers, the shift in paradigm was felt even earlier than this. A significant rise in the weekly wage—a doubling of pay to five dollars a day—and a radical reduction of the cost of the automobile aimed to provide factory workers with enough money to be able to buy their own cars, thus stimulating the potential scope and growth of the automobile market. Aligned with this improvement in workers' conditions was a strategic restructuring of the working week, shortened to between forty and forty-eight hours, and the stability of employment status via the promise of a job for life. By 1922 all three of the traditional anxieties of the worker—concerning money, time and the future—had been addressed, and the resulting employment practices developed into dominant working tenets within the Fordist manufacturing plant. Labour turnover in Ford's manufacturing plants became so negligible that he observed it was not even worth evaluating (Ford and Crowther 1922).

Even with such advances in the working conditions for unskilled labourers, many critics condemned the problematic dynamics of labour within the Taylorist-Fordist factory, none more incisively than Italian communist and political theorist Antonio Gramsci, who famously argued that industrialisation had succeeded 'in making the whole life of the nation revolve around production. Hegemony here is born in the factory' (Gramsci 1999, 285). In one of his many perceptive analyses of the plight of workers in their new, mechanically oriented and 'degrading' roles, Gramsci eruditely deconstructed the ways in which important psychological aspects of the labour process were suppressed so that the body could carry out monotonous, repetitive actions in coordination with the new 'life-draining' technical regime of the conveyor belt and its attendant mechanisms. Gramsci made it clear that he regarded the process of constructing impersonal and standardised mechanistic bodies—bodies that in many ways reflect the status of the objects being produced—as one that began with industrialisation. 'Taylor', Gramsci wrote, 'is in fact expressing with brutal cynicism the purpose of American society—developing in the worker to the highest degree automatic and mechanical attitudes, breaking up the old psycho-physical nexus of qualified professional work, which demands a certain active participation of intelligence, fantasy and initiative on the part of the worker, and reducing productive operations exclusively to the mechanical, physical aspect. But these things, in reality, are not original or novel: they represent simply the most recent phase of a long process that began with industrialism itself' (302).

James Womack, Daniel Jones and Daniel Roos sum up the historical and ongoing importance of Ford's industrial methodologies when they surmise that the auto industry has 'twice in this century . . . changed our most fundamental ideas about how we make things. And how we make things dictates not only how we work but what we buy, how we think, and the way we live' (Womack, Jones and Roos 1990, 11). Most pertinent to our purposes in this book are the ways in which the early industrial factory shaped and composed the industrial workspace, its rhythms of labour, the psychological rupture of alienated labour, the workforce's collective and individuated movements in coordination with automated processes and, in particular, the workers' relationship with the sonic landscape of the factory environment. The dynamics that configure all of these relations were irrevocably changed when the electrically powered moving assembly line was introduced as the automated conductor of workers' rhythms, techniques and agency. In terms of sound and hearing, as the sonic terrain of the worker changed from natural to machine-driven, the continual motion of the conveyor belt became an aural signifier of these transformations. Never allowing for silence or interruption, the industrialised sonic domain—noisy, repetitious and relentless—marked out the territory of the factory and bled outwards onto the street and beyond.

Moreover, just as Taylor's monograph became the industrial handbook of scientific rationalism, helping to shape 'Fordism', so it also preempted the utilisation of sonic media within the factory to assist in ensuring efficient practices and attaining the manufacturing goals set by management. In 1922, as the Fordist doctrine of functional specialisation and the division of labour flourished, and at a time when the workforce had gained some ground and found some satisfaction in their working conditions, a new system of 'Wired Radio' was made available for the industrial space of the automated industrial plant. Created by US Major General George Owen Squier, this technology allowed radio programming to be wired into factories, restaurants, small businesses and individual subscribers' homes. Overcoming the problems of signal loss that were all too regular with radio broadcasts, Wired Radio supplied an endless programme of music over electrical lines, with no commercials or interruptions, for both of which Squier had a known contempt. He also had little patience for the waves of privatisation that had slowed down development of the early telephone industry. In response, he patented his invention in the name of the American public, making the technology legally available for anyone to develop further in the hopes that improved versions would be produced.

As a communications technology, radio is commonly lauded, or indeed denounced, for its capacity to produce a coherent mediated social body through simultaneous long-distance delivery of sonic programming to many listeners. This desire to collapse space and time through technological means has long been a central human ambition (as exemplified by the US military's public development of the Internet). The invention of Wired Radio would go some way towards sonically shaping and moulding this coherent social body, but primarily in the form of the workforce in its concentrated industrial setting. Having rebaptised his technology with a new name under which it was to become famous, Squier would not live long enough to witness the first successful transmission of 'Muzak'[3] into commercial outlets in New York City in 1936. Shortly after the system had been sold to Warner Bros., and then again to William Benton, World War II broke out, and the potential of Muzak to help orchestrate and conduct work rhythms in factories became apparent as it became the 'naturalised' environmental soundtrack of the manufacturing complex.

With the advent of Muzak, the electrically powered arteries that carried music to all parts of the workplace become the sonic equivalent of the electrically powered assembly line that utilises and demarcates each space of the factory. The whole spatiality of the factory building is composed around the productive sequencing of the assembly line, with no section of the industrial manufacturing space left untouched or unmarked by its movement, rhythm and repetition. The factory space with its open planning introduced a new fluid architecture predicated around the assembly line's free-ranging access to mark and touch every space, every subjectivity; its call is to assemble all present around its modal logic of distributed transience. The repetitive processing of this flow engaged the body in a direct relationship with industrial transfer, and it is Muzak that scored the body, orchestrating its part in an extended symphony of staccato manoeuvres.

The architectural form of the cell—so important to Foucault's analysis of the history of the prison in *Discipline and Punish* (where he declares that 'the disciplinary space is always, basically, cellular'; 1975, 143)—was redeployed in the waveform domain by Muzak's temporal ordering of the factory's sonic environment. While the conveyor belt effectively deconstructed the rationale of the monastic cell, with its compartmentalisation of space and time, Muzak sonically contradicted it by mimicking it. In the new musical industryscape, rhythms, spaces and workers were connected by a melodic structuring of time, as the work day was sonically reorganised and categorised via Muzak's harmonious formulas and by the (musical) silences between musical programming (although of course, there is never silence in the factory, except in the case of mechanical failure). Accordingly, the cellular programming of silence became a way to produce temporal, physiological and psychological meaning within the factory and a way to classify the rationale of the sound that preceded or followed it.

Analogously to the spatial-networking capacity of the conveyor belt, the musical wires of the speaker system constructed and rationalised the architecture of the industrial soundscape with a new transcellular order. The speakers at the end of each set of wires became aural portals through which workers entered and exited the overlapping soundtracks of the machinery and the amplified camouflage of the parasitic musical programming. Viral in nature, the epistemic modality of the piped music required the systemic architecture of capital's spatial logic to operate as a carrier. Inversely, with the arrival of Wired Radio the factory finds itself furnished with a new nervous system, a network of cables to propagate and redistribute its sonic load. From this point on, the Muzak sound system possessed the capacity to recompose the factory's sonic spatiality, from one filled with the fractious and chaotic aural by-products of workers and machinery to one made subservient and predictable.

Just as Ford was one of the first to use an electric motor to drive the assembly line, Muzak was the first industrially functional music to be electrically amplified and distributed throughout the work environment. In this way electricity radically respatialised the flow of manufacturing in the Fordist workplace, for the first time allowing a single piece of music to transfer simultaneously into each and every part of its architecture. As such, the spatial code of the industrial sonic landscape was reimagined by Squier as the relationships between the factory's peripheral latitudes and centralised concerns were recomposed by Muzak's radio-controlled pathogens.

AUDIOANALGESIA

The rationalisation of time and space via the logic of capital seems to find its perfect soundtrack in the form of Muzak. Yet writers such as Joseph Lanza have claimed that Muzak in the factory was predicated more on the healing of workers than their organisation by and for capital: 'music was not entertainment but an "audioanalgesia" to kill the pain of urban din' (Lanza 2004, 11–12). In this reading, Muzak in the workplace is posited as a harmonious gesture of empathy from management, a waveformed method of pacifying the body in its new inhuman relation with machinery. There are many problems with this analysis, however, none more telling than the fact that the 'pain' Lanza alludes to was caused not so much by the body being subjected to a new mechanical soundscape as by the requirement of adherence to this new noisy territory. In the overall symphony of the production line, the body is rendered as a numbed note within the staccato rhythms of industry. Muzak would ultimately become the lullaby of the automaton, as the dystopia of random noise was blanketed by the capitalist utopia of repetitive melody. The factories and mills were the places in which these incongruous modes of sonic spatiality would fuse for the first time to produce bodies disciplined against their own natural biorhythms; it was the forced industrial choreography of the work day that necessitated musicality in the soundscape to shape the bodies' new mechanised rhythms and movements. This is the sonic terrain adroitly comprehended by French politician and economist Léon Faucher as that spatiality where the disjunctive power relation of the machine over the body can most easily be perceived. 'Go into a cotton-mill', Faucher wrote. 'Listen to the conversations of the workers and the whistling of the machines. Is there any contrast in the world more afflicting than the regularity and predictability of these mechanical movements, compared with the disorder of ideas and morals, produced by the contact of so many men, women, and children' (quoted in Foucault 1975, 244).

The predictability of mechanical movements became the kinaesthetic cornerstone of industrialisation's relation to the body, structuring and training its operations from the minute employees entered the workplace to the minute they left. According to DeLanda, this process of rendering the organic unpredictability of the human body subservient to the logic and rationale of the machine had begun much earlier. 'The military process of transforming soldiers into machines, as well as related campaigns to organise the management of human bodies (in military hospitals, for instance) generated much knowledge about the body's internal mechanisms', writes DeLanda. 'The "great book of Man-the-machine" was both the blueprint of the human body created by doctors and philosophers and the operating manual for obedient individuals produced by the great Protestant military commanders—among them, Maurice of Nassau, Gustavus Adolphus and Frederick the Great' (DeLanda 1991, 138).

DeLanda maintains that the military-industrial complex had been materialising over centuries of practical and logistical exchange between the civilian economy and the army. As economic and military organisations were transformed and mutated by this continual dialogue, it became clear that military inventions such as Muzak could also aid in the industrial organisation of the mass body of labour to directly support the war effort. The ultimate goal was to mass produce objects that had interchangeable parts, with a labour force that was itself dispensable and—partly through the intervention of music—standardised, a set of precursor techniques that point to the contemporary obsolescence inherent in the production and distribution of music, architecture and objects in general.

Muzak's standardisation of music was the first time in history that an attempt had been made to quantify, categorise and classify waveforms via their functional disposition in an industrial environment. This ordering of frequencies also pertained to the movement of workers' bodies at specific times of day (and night). As such, its scientific rationale emerged at the nexus of industrialised temporality, somatic engineering and architectural routine. When elucidating the founding principles of the prison system, Foucault also registers the strategies and disciplinary techniques that would subsequently be transferred into the industrial realm, making it clear how the body was subjugated to political, social and economic methods of utility. 'The human body was entering a machinery of power that explores it, breaks it down and rearranges it', wrote Foucault. 'A "political anatomy", which was also a "mechanics of power", was being born; it defined how one may have a hold over others' bodies, not only so that they may do what one wishes, but so that they may operate as one wishes, with the techniques, the speed and the efficiency that one determines. Thus discipline produces subjected and practised bodies, "docile' bodies"' (Foucault 1975, 138).

Foucault's 'docile body' recalls Lanza's conception of 'audioanalgesia' and the composition of the numbed body it implies. For Lanza inadvertently hits upon the notion of the sedated body here—the body that wishes to kill the pain by having its industrial organs soothed by music. This early-industrial body is laid bare, vulnerable in its newly composed state, awaiting its sonic anaesthesia so that the operation can begin. What is interesting here is that the process Ford applied to mass production and consumption—namely, his system of standardisation, of manufacturing techniques and components—was also applied to the soundscape in order to achieve a spatiality characterised by modular interchangeability and repetition. Serial numbers were inscribed on parts of objects such as those belonging to guns and cars so that they could be classified and easily swapped out or repaired. And similarly, to produce waveformed Muzakal objects, the working day was broken down into fifteen-minute segments that were subsequently serialised and categorised in much the same way as parts of an automobile so that each Muzakal module could be broken down, replaced or repaired if it was deemed dysfunctional—that is, if it stimulated the workers too much or too little.

As the conveyor belt repetitively delivered objects to be worked on, one task at a time, so Muzak conveyed sonic parts, one track after another, to complete a full fifteen-minute sonic object that worked on the employee. The process was almost an inverse of the conveyor belt—an early waveformed heterotopia[4] in which every part of Muzak's scientific rationale was operative in any single song, as each fulfilled a function in relation to all the other songs played in that segment, hour or day. Simultaneously, on the fringes and at the nexus of music, industry and the social sciences, Muzak can be heard as an audiotopia—the sonic equivalent of Foucault's ultimate example of a heterotopia in the form of the contradictory mirror. Muzak reflected music's power to unite, to motivate, to shape patterns of economic and social behaviour and to compose somatic rhythms, yet it had no social meaning outside of the workplace. This rendered it a conflicted waveformed spatiality with a paradoxical identity: expressing both unity among employees and the enforcement of industrial work rationale. Muzak was both utopian painkiller and dystopian agent of embodied discipline.

By attempting to quantify, temporalise and rationalise the sonic sphere, Muzak aimed to make waveformed territories knowable, controllable, perceivable and available for the purposes of indexing and stimulating human actions. Within these newly defined test sites that constituted the industrial workplace, the body became the object of scrutiny, of the stimulation of affect and, ultimately, of control. This is the body Foucault is speaking of when he proposes that 'in becoming the target for new mechanisms of power, the body is offered up to new forms of knowledge. It is the body of exercise, rather than of speculative physics; a body manipulated by authority, rather than imbued with animal spirits; a body of useful training and not of rational mechanics' (Foucault 1975, 155). It is the useful body of the worker, alienated from the objects he produces, little more than 'a mechanical part incorporated into a mechanical system' (Lukács 2002, 89), that finds itself situated in a recomposed sonic environment of new collective rhythms and coordinated movements. The somatic organisation of the worker in this environment echoes the agenda of the production line. 'Already preexisting and self-sufficient', Lukács describes the context, 'it functions independently of him and he has to conform to its laws whether he likes it or not' (ibid.). With Muzak ensuring that cartographies of these new waveformed spaces were catalogued in the libraries of knowledge that aimed to perceive, predict and ultimately 'know' the body's behaviours, the somatic was recomposed into unprecedented choreographies of mass industrial movements. Through the Muzakal filter, music was arranged and applied to lure the body into obedience to disciplinary methods that would further rupture the industrial subject's capacity to act independently within the workplace.

TRANSNATIONAL HARMONIES

Over time, Muzak's multiplexing process (multiple analogue or digital signals converged into a solitary signal and transmitted over a communication channel) developed into cable technology, the kind that today delivers television into millions of homes. As this communications technique became ubiquitous and helped pave the way for the notion of the global village[5], so Fordist modes of standardisation foreshadowed practices that would shape later waves of economic globalisation. As Squier dreamed of being able to amplify music into workplaces and homes all over the United States and beyond, so Ford helped usher in a set of economic principles that defined an era of transnational exchange of goods and people. Through the implementation of information systems such as cable and the industrial processes of mass production that were inaugurated by Fordism, spatial notions of what constituted a nation, a territory and a community were changed irrevocably. Along these lines, American sociologist Saskia Sassen points out that 'globalization—as illustrated by the space economy of advanced information industries—denationalizes national territory. This denationalization, which to a large extent materializes in large cities, has become legitimate for capital and has indeed been imbued with positive value by many government elites and their economic advisers' (Sassen 1998, xxviii).

By redefining the spatial dynamics of the waveformed terrain and physical landscape alike, industrialisation and its concomitant information-based and production-based technologies not only abstracted relationships with the body but also altered its relation to the frequencies in which it was immersed. For the first time, sound—with the covert intent of organising employees' working rhythms in space—was amplified on a mass scale, anticipating the ways in which bodies would be shifted en masse via later globalist-inspired treaties such as the North American Free Trade Agreement.

Karl Marx and Friedrich Engels were among the first to anticipate that capitalism was, in the long term, inherently destined to become globally expansive[6]. 'All fixed, fast-frozen relations', they wrote, 'with their train of ancient and venerable prejudices and opinions, are swept away, all new-formed ones become antiquated before they ossify. . . . The need of a constantly expanding market for its products chases the bourgeoisie over the whole surface of the globe. It must nestle everywhere, settle everywhere, establish connexions everywhere' (Marx and Engels 2007, 18; first published in 1848).

The final sentence here could almost be speaking directly about Muzak and its creator's wish to correlate a mass population with the programming of a 'scientifically' manufactured soundscape. The range of musical programs on offer from Muzak—engineered soundscapes for the workplace, for the home, for leisure time—required constant connection and validation from its listeners in order to expand and proliferate. For globalisation theorists such as Sassen, 'globalisation is a process that generates contradictory spaces, characterised by contestation, internal differentiation, continuous border crossings' (Sassen 1998, xxxiv). This notion of connected spaces that are contradictory in nature again resonates with Muzak's sonic engineering of the factory—the working blueprint of the industrially globalised world—and its attempt to suppress noise and dissonance through melody and harmony.

Muzak proposed to negate the random, chaotic and disturbing nature of the industrial sonic terrain, advocating instead for military-based technologies that could deliver new malleable sonic architectures, frequency-based formulas that could be arranged to work in any geographical context and orchestrated to bring ordered collective reasoning and attuned compatibility to any social, leisure or working situation. The acoustic realisation of Muzak in an industrial setting produced new contradictory compositions made up of oppositional dynamics, those of noise (from the machinery) and melody (from the Muzak). For the first time, with the use of electricity, the juxtaposing aesthetics that we so readily accept today over a range of musical genres—those of harmoniously organised sound (signifying the connected) and randomly discordant cacophonies (signifying the alienated)—entered into one and the same score within the factory, with the worker's body becoming the anatomical mixer through which all frequencies were channelled and amplified.

MASTERS OF DISCIPLINE

In the late nineteenth century, it was in cathedrals and churches that the mass social body would congregate. Within such architectures, religious practice, instruction and communal ties were expressed and validated by groups of spiritual believers. As well as celebrating the visible signifiers of belief—the cross, the images of stained glass, scripture—perhaps more importantly these spaces collectively reproduced frequency-based signifiers in song, organ recitals and oration, the bass notes emitted by the organ causing low, rolling infrasonic frequencies creating a sense of awe and trepidation. With the onset of industrialisation, mills and factories offered new architectural locations within which the mass social body would congregate, and the waveformed techniques employed in the organisation of large numbers of people were transferred from the place of worship to the place of work. Discussing this assimilation by industry of the methods used by religion to spatialise and territorialise the sonic domain, Lanza states that 'modern capitalism instigated an ecclesiastical rift. If background music was good enough to orchestrate the houses of God, why not the houses of commerce?' (Lanza 2004, 11–12).

As ecclesiastical values were relegated to the background of everyday life by modern capitalism, then how could music help prioritise the psychological requirements of the houses of commerce over the needs of the houses of God? As a space for the dissemination of waveforms, the church—previously socially sanctioned as the most significant producer of sound in any city, town or village—lost its architectural dominance over the sonic realm to industry. Suddenly the frequency-based blueprints that had been composed by religion were now being orchestrated by the socioeconomic dictates of capital. Of particular interest here are those waveformed techniques aimed at psychologically influencing and manipulating the mass social body: practices designed to unite the individual into a collective of belief while also alienating them by negating the sonic space necessary for individual expression. Workers now had to react to the demands of a foreman rather than follow the cues that had formerly prompted them to collectively sing in the field or church. Now the industrial worker had reason to question the rural and religious belief patterns that had formerly defined the architectures of his existence. Accordingly, his relationships with the traditional waveformed dynamics of public and work space also became uncertain.

In the factory a whole new world of spatialities, temporal modalities, sonic domains and social relations was emerging, originating in part from previous repetitive arrangements of the working day developed by religious systems. As Foucault recalls, 'The *time-table* is an old inheritance. The strict model was no doubt suggested by the monastic communities. It soon spread. Its three great methods—establish rhythms, impose particular occupations, regulate the cycles of repetition—were soon to be found in schools, workshops and hospitals. For centuries, the religious orders had been masters of discipline: they were the specialists of time, the great technicians of rhythm and regular activities' (Foucault 1975, 149–50; emphasis original).

As the new dictator of social- and labour-based rhythms, Fordist capitalism began to compete with religious organisations for lifelong affiliation—from a church and a set of beliefs for life to a factory and a job for life. In place of righteous assemblies bringing the community together, unions brought employees together to empower them against unjust forms of oppression in the workplace. Most pertinently for our purposes here, this shift from the preacher to the foreman was paralleled by a shift from the promised omnipresent voice of God that came from above to the omnipresent music that encompassed all space at all times, emanating as it did from the factory's network of wired speakers. The spatial construct of mounting speakers above head height mimicked the religious longitudinal ordering of waveforms in which the 'voice of God' comes from 'Heaven on high', a preordained construct that finds its frequency-based expression in Psalm 18:13: 'The Lord thundered from heaven; the voice of the Most High resounded' (New International Version).

It is no coincidence that the frequency-based power relations enacted between the church and its attendant subjects—the enormous pipes of the organ reinforcing psychologies of domination and submission—should be reimagined in the factory. As the architecture of the cathedral functions in part to organise its members into collective patterns of associated behaviours and actions by producing a sense of threat through the sheer size of the building within which these commands are transmitted, so the factory dominates its inhabitants' comportment. Foucault tells us that 'the factory was explicitly compared with the monastery, the fortress, a walled town' (Foucault 1975, 142); yet this is to disregard the divergent sonic architectures of the two structures. The factory at this point in time was possibly the noisiest man-made environment outside of war's turbulent cacophony, its frequency-based topography in no way comparable to that of the hushed monastery. The cathedral engaged in a sonic politics radically different from that of the monastery, and its dynamics were to be echoed by industrialists when they filled the vast overhead spaces of the factories with waveformed authority. Accordingly, the Muzak emanating from the space formerly reserved for the voice of God signified the presence of a new kind of power that, while disinvested from spiritual aspirations, was very much invested in the frequency-based techniques of influence and manipulation that religion had mastered.

The spatial sanctity of the cathedral, and more particularly of the area above head height, was and still is reserved for surveillance and sonic expression from a ubiquitous intelligence. And at the beginning of the twentieth century, this dynamic was strategically deployed in the churches of capitalism—the factories. Transmitting supposed psychological messages of 'solace' and 'humanity' to the workers as they toiled among machinery, the networked speakers also subtly reminded each employee of the existence of a higher power capable of amplifying its presence through each and every square inch of the factory, from wall to wall, from ceiling to floor. This set of architectural relations announces the presence of an unseen intelligence, a waveformed phenomena that can extend itself to all places at all times and that can choose, at its own discretion, to provide an audible sanctuary or sonic battleground. It is precisely this composition of frequency-based presence that creates a sense of self-surveillance in the industrial subject, as he is constantly made aware of his passive position—his capacity to be recorded and for his agency to be drowned out—within the factory's envelope of speaker-driven power relations. By its very nature, the transmission network with its wired aural arteries endowed one with what was formerly regarded as an otherworldly power—namely, the ability to extend one's presence into multiple spaces at the same point in time.

MASS NEURAL NETWORK

In the great production houses of industry, the ever-shifting terrain of the workers' emotional and psychological status became objectified as a valid subject of study. Research and testing was undertaken into the cognitive dynamics and behavioural patterns of the worker as he undertook repetitive tasks, and systematic taxonomies of interpersonal relationships were created as an aid to promote harmonious functionality. In short, the psychological landscape of the worker was identified as a locus of natural energy that needed to be catalogued and understood so that factory owners might best profit from its newly engineered status and potential. The emotional reserves of employees came to be regarded as a kind of somatic fuel to be extracted and redistributed according to the rhythms and efficiency dictates of the production line—and it was the wired Muzak network that would serve as its transmission system.

In the 1920s factory, the first attempt to construct a mass neural network of productivity via the influencing strategies of Muzak manifested, with each mind becoming a functional point of reference for ultimate industrial efficiency. In the acoustic laboratory of the Fordist factory, epistemological strategies to locate, chart and manipulate a mass psychology staked out the first sonically colonising markers of a somatic industrial rhythm. Squier's networked creation had produced an aural geography in the factory in which the topography of the worker's mind was the territory under surveillance. It was through this newly knowable psychological spatiality that the sequencing of somatic tempo could be programmed and the cadence of productivity ratios mastered.

As the emotional and psychological behaviours of the early-industrial body during periods of pressure, duress and calm became objects of scientific study, a myriad of laws, theories and tests were advanced under the moniker of 'industrial psychology' in order to prove that the human mind could be influenced and manipulated within the workplace and that as a result, the efficiency rates of the body could be improved. The notion of employing music as a stimulus within the workplace had been proposed before many of these theories were drafted. Two years before Squier's speaker technology became commercially viable in 1922, Thomas Edison and his National Phonograph Company had been busy researching the heuristic application of music. As early as 1915, Edison had developed a number of ideas about the power of music over both the individual and the mass social body. He carried out experiments to test whether piped-in music could cover up or negate specific frequency ratios produced by the heavy industrial machinery of the factory and to ascertain whether the workers' morals and work drives could be positively or negatively affected. Buoyed by his early findings, Edison recognised the potential of music to affect listener's dispositions and became deeply interested in its capacity to direct emotions and actions within the workplace.

Further tests were undertaken by Edison to produce 'sonic camouflage' for the factory to cover the drones and mechanical mutterings of its machinery. The trials carried out in the newly compartmentalised workspaces of the factory were not successful, however, owing to the inadequate signal strength of early transmission and loudspeaker technologies. Nonetheless, it is in these experiments that an initial drive to psychologically manipulate a distributed workforce via a networked system of electrically powered speakers can be perceived. Edison's initial investigations into the controlled psychological manipulation of the mass body and its repetitious routines were the first of their kind, providing the cursory acoustic blueprints for the functional use of sound within an industrial architecture. But while Edison had made it possible to think about simultaneously emitting music in a plethora of diverse spaces, technology had not caught up. Only when Muzak's sound systems came into being could factory owners engineer the same acoustic profile in multiple spaces over any given duration of time.

Edison was keen to motivate his employees and coworkers to draw the utmost functionality and utilitarian application out of any of his inventions. In light of his desire to market the multifaceted effects of music to a range of industrial, social and cultural groupings, in 1920, under the auspices of the National Phonograph Company, he employed Walter Van Dyke Bingham, an assistant professor of applied psychology at the Carnegie Institute of Technology, who would later go on to become an industrial psychologist. Bingham was contracted to use the company's burgeoning archive of phonographic recordings to study and quantify the 'effects of music', in a research programme defined by the three key criteria of song selection, mood change and the influences of music on muscular activity. Bingham's earlier psychological and philosophical research hinged on the problem of why certain tonal arrangements constituted melodic unity and secondarily, how such melodic stimuli (when achieved) subsequently influenced a human's motor movements. His interest in industrial and somatic motor functionality between 1910 and 1920 is of no little significance, according to writers such as Eleanor Selfridge-Field, who has documented the history of efforts to rationalise somatic movement in the industrial workspace. Tracing out the ways in which Bingham harnessed the potential of the motor and reconfigured often clumsy industrial processes into a fluid spatiality of perpetual monophonic flow, Selfridge-Field makes clear how these movements were analogously mapped onto the nervous system of the human body, citing Bingham's conclusion that the

> motor theory of melody makes possible an unambiguous statement of the nature of melodic 'relationship'. Two or more tones are felt to be 'related' when there is community of organised response. . . . The origin of . . . feelings of 'relationship' . . . [may be attributed to] two main forces. . . . The first of these, the phenomenon of consonance, is native. . . . But although the basis for consonance inheres in the inborn structure of the nervous system and the acoustical properties of vibrating bodies, nevertheless it is a commonplace of musical history and observation that these same native tendencies are subject to tremendous modification in the course of experience. (Selfridge-Field 1997, 293)

Bingham positions the resonating body's distributed sensorium at the vinculum of scientific, phenomenological, musical and industrial discourses. The 'mood tests' he employed consisted of collated and accumulated charts and documents of how his subject's moods altered as they listened to a programme of Edison's musical recordings. In a progress report to Edison dated 1 February 1921, Bingham expresses his hope that his research will produce 'new information about the power of music over men's minds and moods' (Selfridge-Field 1997, 297). On 13 October 1920, he had announced that a prize would be awarded to any researcher who undertook 'meritorious' investigation into one of the following 'appropriate subjects':

1. Classification of musical selections according to their psychological effects.
2. Individual differences in musical sensitivity.
3. Types of listeners.
4. Validity of introspection in studying affective responses to music.
5. Modification of moods by music.
6. Effects of familiarity and repetition: emotional durability of various types of selections.
7. Effects of contrasting types of music on muscular activity.
8. An experimental study of music as an aid in synchronizing routine factory operations. (295–96)

Dissatisfied with the way Bingham's research was heading (the fact that the results of the subject's responses in the tests could not be directly attributed to helping sell specific recordings in the Edison catalogue), a company vice president by the name of William Maxwell took it upon himself to create a 'Mood Change Chart' in 1921 to remedy this apparent utilitarian deficit. This single-page chart posited seven simple questions for the participating test subject, including, 'As a result of the test, what were your most noticeable mood changes? (Serious to gay, gay to serious, worried to carefree, nervous to composed, etc.)' and 'Please comment on manner in which mood changes occurred' (Maxwell 1921). The Maxwell 'Mood Tests' were taken seriously enough at the time to be conducted at Yale University. The analysis of the sonic investigations was, however, reported somewhat fancifully by the *New York Sun*, which Selfridge-Field quotes as predicting that 'music may become useful in treating human maladies'.

> 'The day may come, it is predicted at Yale, when pneumonia will be treated not only with open windows and malted milk but by a few disks of dreamy waltz music. If a man breaks his arm and is restless, a battle march or possibly a line of comic opera may be fed out to him after each meal'. A further piece on the Yale experiment, which appear [sic] in the *Journal Courier* on the 22nd, elaborated on this idea by stating that 'the principal effort of the tests was to determine what kinds of music may be applied in treating neurotic patients'. (*New York Sun* quoted in Selfridge-Field 1997, 300)

In the 1920s, the drive to comprehend, classify and navigate the psychology of those pursuing both leisure and work activities had found a firm foothold in socially scientific research. That said, for economic reasons it was the role of the worker that most interested those conducting the tests and surveys. And no longer were the presence, movements and rhythms of the physical body the only characteristics that management cared to address and order. From a psychological perspective, the managerial desire to alter the mood of a worker was analogous to the way that a supervisor might speed up or slow down a conveyor belt or a phonograph. This desire to boost a company's psychological leverage over its workforce was enveloped in the seemingly innocuous language of 'boosting morale'. It posited phenomena such as music as a collective experiential stimulant to shape and improve behaviour, with the underlying promise of emotional amelioration—as if the act of listening was more to the workers' advantage than the company's.

Redefining the temporality and spatiality of the factory, Muzak can be perceived—with reference to the findings of the 'Hawthorne studies'[7] and, by extension, through the human-relations movement[8]—as being emblematic of the continual expansion of management's presence via sound. In this sense, Muzak became an industrialised inversion of Jeremy Bentham's Panopticon[9]—the networked speakers being dispersed, amplifying and inhabiting all spaces. They proposed a diffused ideology rather than the centralised arrangement of the Panopticon that observed the surrounding cellular spatiality of the prison. Yet the behaviours that result from being subjected to these systems—if read through the conclusions of the Hawthorne studies—are similar. What is most important here is the act of redefining psychological spatiality—that of the prison in the case of the Panopticon and the factory in the case of Muzak. The principal factor of technological ownership and the capacity to extend one's presence and influence into another spatiality at any given time is what redefined the power relationship between guard and prisoner, management and worker. In the prison setting, the Panopticon threatened to extend the guard's vision into a direct relationship with the activities of the prisoner, while in the factory the speaker system extended recordings into the spatiality of the workplace and into the emotional terrain of the worker's consciousness.

A number of influential investigations that attempted to rationalise and predict the physiological and psychological cartography of the worker had also been undertaken before Muzak was invented. Formatted independently of one another, in the 1880s two similar theories became conjointly known as the James-Lange theory[10]. It proposed that emotions are triggered by physiological changes in the body manifested via experiences in the world. Three decades later, the Cannon-Bard theory of emotion[11] would reverse this order of priority. But among the multitude of theories of the late nineteenth and early twentieth centuries that attempted to comprehend, systematise and organise the behaviours of the mind, the psychological approach most influential for the development of Muzak was the law postulated in 1908 by psychologists Robert M. Yerkes and John Dillingham Dodson. The Yerkes-Dodson law proposed that there is an observable relationship between levels of arousal and performance, and that a worker, for example, would be more productive if they are psychologically and/or physiologically aroused. The levels of stimulus must be controlled, however, because too much arousal would become detrimental to the worker's concentration. The theorem went further in stating that different types of activities required different specific intensities of arousal in order to ensure optimal task efficiency. According to the Yerkes-Dodson law, those activities requiring intellectual prowess demanded a lower arousal intensity—allowing the worker to concentrate better—while repetitive and physically tiring assignments demanded that a worker be stimulated to higher degrees of arousal in order to increase his motivation.

Following the dictates of this law, during the 1930s and 1940s Muzak engineers indexed all actions, emotions and human relations into a musical framework, a technique that culminated in their elaborate programming of fifteen-minute blocks of music known as *stimulus progression*. This new form of cataloguing and indexing sound was implemented by Harold Burris-Meyer and Richard L. Cardinell but was masterminded by Muzak executive Don O'Neill, who had toyed with the idea since joining the company in 1936. Premiering in the late 1940s, stimulus progression was a method of organising music according to the ascending curve that worked counter to the industrial efficiency curve (also called the average worker's *fatigue curve*)[12], as stipulated by Yerkes and Dodson.

Subdued songs, progressing to more stimulating songs, in fifteen-minute sequences (with silences between transmissions of thirty seconds to fifteen minutes long) throughout the average workday yielded better worker efficiency and productivity than did random musical programming. Programs were soon tailored to workers' mood swings and peak periods, as measured on a Muzak mood-rating scale, ranging from 'gloomy—minus three' to 'ecstatic—plus eight'. Songs were thus categorised by Muzak according to their 'stimulus' capacity, a measure that incorporated rhythm and tempo analysis, types of instrumentation and orchestra size, enabling a full classification of any given song's propensity to encourage optimum labour throughout the workday. Constructing its ratio in relation to the rhythm of the human heartbeat, stimulus progression's median tempo was set at seventy-two beats per minute—the cadence of the average human heart at rest. With this new formula in hand and in ear, the early Muzak engineers were ready to territorialise not just the sonic realm of the industrial workplace but any waveformed environment imaginable.

During the unimaginably brutal sound clashes of World War II, Muzak was utilised to instigate a sense of calm in English munitions factories, environments that were regularly enveloped in a cacophony of explosions and sirens. While Muzak was employed to calm workers so that they could work efficiently, the German Army implemented psychological sonic weapons into the design of their Stuka bombers, which were equipped with sirens (activated at the beginning of a nosedive towards a target). The majority of the aircrafts' bombs missed their targets and produced only a roar, meaning that the Stukas functioned as a psychological weapon as much as a physically destructive one; their aim, of course, being to generate collective fear in large groups of people via frequency-based transmission technologies.

Muzak engineers, with very different intentions than the Stuka sound designers, were, however, researching similar waveformed affect. They wanted to know what phenomena made a group oscillate at the same frequency. This is a question still being asked today, as the quest to regulate the activities of a mass body into the more-manageable rhythms of a singular entity remains fundamental to organisations such as governments, corporations and militaries. To harmonise the activities of humanity is to rid men and women of their chaotic mind-sets, irregular habits and disobedient social behaviours in order to choreograph and entrain each person with a set of socioscientifically implemented motivations. Making all minds think, and all bodies move, in unison was and still is a dream of those who wish to distribute and bind the collective social body into the inescapable arrangements of capital, leisure or conflict. In the factories of the twentieth century we can hear how the industrial elite standardised the irregular (every)body into a repetitive and replicable production cell by harnessing music to optimise economic outcomes.

The attempt to create mass psychological conditions via organised sound during World War II was not only heard in the factory or from the Stuka-strewn skies but also via radio propaganda issued by both British and German governments on a regular basis to try and deceive one another and to increase levels of national belief and camaraderie. Touching on the capacity of frequencies to induce cooperation and to facilitate the compression of multichannel activities into a singular rhythm, DeLanda posits that 'almost any population whose individual members oscillate or pulsate is capable of reaching a singularity and thus to begin oscillating in a synchronized way. When this singularity is actualized and the rhythms of the whole population "entrain", its constituent individuals acquire a natural esprit de corps. This "team spirit" allows them to behave as if they were a single organism' (De-Landa 1991, 64).

Although here DeLanda is analysing the capacity of military groups to move rhythmically and to believe as a unit, the principles he identifies also furnish industrial modalities of orchestrating an esprit de corps. For it is in the sonic domain that we can hear such strategies and techniques bleed noisily between military and industrial bodies of thought. And, as we shall see in subsequent chapters, it is these early-twentieth-century transmissions from the military-industrial complex that will gain in significance as they are re-engineered and find new expression in later symbiotic dynamics such as the military-entertainment complex's organisation of space, bodies and time in Guantánamo Bay.

DISCONNECTING THE GLOBAL VILLAGE

The overarching argument of this book is that, since 1922 and the inception of Wired Radio, frequency-based strategies have been used to disconnect and alienate individuals, families and groups from 'belonging' or 'relating' to their former social networks, architectural contexts and sociocultural affiliations. For thousands of years music has been understood to have the capacity to bring people together to dance, sing and pray. Given that we commonly acknowledge music's capacity to unite humans in a wide array of endeavours, it is logical to conceive that such an effective and influential medium could be instrumentalised for operations of a less-convivial nature. As we have seen, the Muzak piped into modernism's industrial factories could be interpreted as being of the former kind, as a source of camaraderie or consolation. But this would be a surface reading neglecting the crucial (un)social elements of Muzak, as it introduced a further silencing of the workforce so that there was less vocal communication between employees. Previously, in the agricultural workplace of the field in the United States, associated song, which told of pain and emancipation through religion and death, was commonplace (Rosenbaum, 2007), but in the factory such reflexive expression was prohibited, along with any songs that could possibly incite revolt or collective disharmony. The new machines were more important to the factory owners than were the health and welfare of their operators. More often than not, any form of song created by employees was viewed as a diversionary interference in the work process. The frequencies pumped into the factories could, then, be discerned as the first melodies of alienation, as Muzak came to displace the previously composed sonic space redolent with storytelling, complaint, laughter, collective dissonance and idle chatter—the audible components of relationship building and social cohesion that are commonplace within groups undertaking long repetitive tasks.

As the formulaic intentions of stimulus progression were never made clear to the workers, its covert nature renders Muzak an early viral form of sonic modality. It disassociated workers from their architectural surroundings and from the internal and exteriorised acoustic markers that might have given them a sense of progression through the day or night. By alienating employees from the sonic reference points that would otherwise be encountered on a daily basis, Muzak helped compose a disconnected and autonomous set of working conditions, the success of which relied solely on a fully engaged and infected relationship between the worker and their associated machinery. As employees could no longer directly relate to the sounds made by the workplace apparatus and were isolated from sounds coming from outside the factory, they were instead each individually cocooned by the Muzak, which attempted to suppress all of those factors that would aid in orienting the worker and giving them a sense of time. Lanza recognises this attempt to estrange factory employees from their everyday existence outside of the workplace when he writes that 'if Taylorism could monitor the time-lag between clerks reaching for pencils and their marking papers, sound engineers could likewise manufacture their version of the optimum work womb' (Lanza 2004, 27).

With the suppression of external and internal sonic markers of temporality came a new dependency on the factory owners to orchestrate the signalling of a break, lunchtime or the end of the working day and a further loss of independence for the employee. The machinery meanwhile reinforced the worker's sense of disorientation by forcing him to move in an automated fashion and to keep time with its unthinking synchronicity. The capacity of the machine not only to do work itself but also to entrain the somatic workforce meant that the mechanical robot became the phantasm of the industrial factory's production line. Muzak, meanwhile, soundtracked the sonic fantasy in which the robots danced with capital, albeit still, at this point in history, with workers as chaperones—third wheels who did not wish to take part but whose heads were filled with the melodies from the speakers above, with little choice as their hands were forced by the choreography of progress.

As we have seen, in Muzak, (multiplexing) technology that would later help bring about the collective notion of the global village was actively applied to alienate employees both from their fellow workers and from their architectural sense of space and place. Supressing conversation between employees not only prevented idle chatter, it also prohibited voices of dissent from spreading and propagating. This sonic sphere was therefore the first instance of an electrically powered domain functioning between the intent of mass communication and the discontent of estrangement. As noted in the introduction, music has always contained this dual potential, but never had it previously harboured such infectious promise—Muzak possessing the capacity to simultaneously affect mass groupings of people in divergent geographical locations, from the dwelling to the workplace.

The introduction of Muzak into the citadel of spatially controlled repetitive-labour processes was endorsed and ratified at a time in history when the sonic territory of the workplace was relatively unexplored, uncharted and politically innocent. In 1922 there were few directives, regulations or laws that addressed the sonic landscape and what it meant psychologically, physiologically, politically or legally for the body to exist within it. Squier's 'Tayloresque' strategy of connecting each worker in the sonic landscape via the omnipresence of musical cycles formed part of an overarching objective: to compose a comprehensive schema of repetitive actions, productions and payments within the factory that would subsequently extend into a global context of distribution and marketing. Such a viral coordination and classification of time and space, from the factory to worldwide networks of expanding industrial capitalism, is the embodiment of Foucault's notion of how the micropolitics of the local are transplanted and etched onto the logic board of global capitalism.

The Muzak headquarters—from which the networks stemmed and from which the programming was carried out—were secreted away from the workers, inaccessible to them. When listening to music over the speakers, the factory employees could not change any of the music's settings, slow it down, manipulate or destroy it, because for the first time ever they did not even know from where the music was being transmitted. Just as the machinery in the factory alienated the workers from their labour, so the networked music separated the body from its architectural surroundings by estranging the workers from the transmission hub where the music originated. The Muzakal bunker spawned transmission networks that directed information and influenced patterns of behaviour, cementing a perception of geographical and psychological control that amplified the ideological dominance of the factory owner. It was precisely this detachment—being made aware that one existed on the other side of the network's interface—that rendered the worker a passive subject and placed the mass industrial body in the disadvantageous position of having no agency within the composition, distribution or destruction of (waveformed) information.

A FLESHY CADENCE OF THE ANTENNA-BODY

Within the newly composed sonic domains of industrialisation the emergence of the antenna-body into an industrial context can be heard. Animated by electricity, the antenna-body alternated between the promise of future systems of information exchange and the utility of available strategies to seduce workers into the embrace of the machine. Employees of the industrial factories, then, were the first bodies to engender such antithetical and divergent waveformed ontologies, both receiving the programming that would create silence between them and concurrently transmitting a prescient signal of technology's prowess in collapsing space and time through the networked logic of the sound system. The early effects of the wired public speaker address system—its resonating development of the antenna-body into a fleshy industrial operator dependent on the sonic as much as the visual—was to echo far and wide. The electrical amplification of sound in private, public and interstitial spaces would go on to have a significant impact on the waveformed terrain of the twentieth century. The distributed agency of electronically aided oscillation was emerging.

Back in the factory, the antenna-body found itself in a sociopolitical interstice between the promise of technological emancipation from the confines of geography and temporality and the reality of being bound by the rhythmical attrition of industrialised production targets. Spatially displaced and mechanically resituated within Muzak's categorised fifteen-minute segments, this body was for the first time disciplined by electrically propagated waveforms. It was observed, documented and analysed within a scientifically conceived factory, which joined the laboratory as one of the significant modern sites for examining the physiological and psychological activities of the body. While learning a new sonically spatialised discourse arranged through the vocabulary of timbre, rhythm and instrumentation, the antenna-body was forced to perceive its new role as one that was as transposable as it was dependent.

Occupying the areas between silence and mechanically occurring sound, Muzak functioned by synthesising the loss of verbal language with the creation of a definitive waveformed terrain and in the process rendered an alternative reading of sonic spatiality—an unconscious topography of frequencies that can be designated *thirdsound*. In the early industrial days of Muzak could be heard a sonic technique that was designed to exist between the waveformed spatialities of noise and silence. In the formative years of the industrialised workplace, silence was feared, given that it was ultimately understood to be a signifier of stasis, an unwanted interlude in the industrial symphony of constant mechanised movements. Silence meant a disruption on the line and thus a breakdown in production. As a progeny of the military and entertainment industries, Muzak came to life by adopting the rhythmical nature of the production line's needs. It grew by channelling workers' and consumers' desires and left us equating silence with death. Lanza describes this industrial fear of silence when he remarks that 'as the Industrial Revolution introduced the internal combustion engine's roar and the drone of generators, ventilation systems, riveting pistons, and low-frequency electrical lighting, silence became an unwelcome anomaly when it existed at all' (Lanza 2004, 11). With hindsight, Muzak opened up the dualistic presumptions about industrially occurring sound and silence by offering new ways of thinking about how spatiality, psychology and pressure within the workplace had been orchestrated by waveforms. As silence gives meaning to the sonic, so Muzak gives impetus to the notion of thirdsound and in doing so invites us to renegotiate our embodied relationship with both one and the other.

NOTES

1. A phase of the Industrial Revolution that started in the mid-nineteenth century with the advent of new transportation systems such as steam ships and railways and with major advances in the steel, chemical and, most importantly for this study, electrical industries.

2. Frederick Winslow Taylor conducted the first time study in the 1880s. Time and motion studies precisely measure all movements within the workplace by analys-

ing the management and machinery of industrialisation in order to work out how the entire working system could function more efficiently.

3. Major General George Owen Squier renamed the corporation according to his interest in the invention of the Kodak company's name. In turn he created a portmanteau of the words *music* and *Kodak* to form the name *Muzak*.

4. Michel Foucault's original conceptualisation of *heterotopia* was of a space that exists between real space and utopian space. As such, it is an other space, functioning beyond hegemonic ordering—a space that is at once real and imaginary, mental and material, here and there. Examples include the mirror, the asylum where deviant bodies are placed and the ship. In the short, posthumously published paper 'Of Other Spaces: Utopias and Heterotopias' (2002), Foucault drew up blueprints for what we now call *heterology*, the study of the other. Heterology has subsequently been used as a tool of critical analysis by a range of scholars and practitioners, from cinema, poetry and urban studies to contemporary art and cartography. Architect Georges Teyssot (Teyssot et al. 1977) has advocated the potential of heterotopia for communicating the sociopolitical dynamics of the built environment, in doing so disclosing the notion's subversive spatial critique. The notion of heterotopia is useful for this study as it suggests that the identity, sociopolitical construction and power relations of any waveformed space are in a state of constant mutation and renegotiation.

5. The term *global village* was first coined by Marshall McLuhan in *The Gutenberg Galaxy: The Making of Typographic Man* (1962), where he outlines how the geography of the globe has been shrunk to the dimensions of a village by electronic communications technologies and their capacity to transfer information from any given point to another at great speeds. He proposes that with such contractions of physical space, humans have an increased sense of responsibility to a wider sense of community than ever before. Lanza states that through Muzak, Major General George Owen Squier 'helped usher in the "global village" by concocting a hook-up system capable of compressing vast areas of time and space previously isolated by geography' (Lanza 2004, 23).

6. Marx and Engels never used the word *globalisation*, which is a more recent term. Rather, they predicted that capitalism as a global expansionist system would transform societies, nation-states and, eventually, the entire world.

7. After Thomas Edison stopped funding studies into the effects of music, the Hawthorne Works (under the aegis of the Western Electric Company) in Cicero, Illinois, began conducting the first of five studies that started in 1924 and ended in 1932. They began by altering workplace stimuli such as music and light to determine whether their employees would be more productive with increased or decreased amounts of either. It was found that production levels did in fact rise when light levels were changed either way. It was concluded that, rather than the manipulation of the environment being the decisive factor, it was employees' own cognizance of the fact that they were being observed that increased efficiency. This effect—which Henry A. Landsberger (1958) dubbed the 'Hawthorne effect' after analysing the results of the study over thirty years later—is now used to describe any short-term increase of productivity.

8. Emanating from the Hawthorne studies, the human-relations movement was an American school of sociology based largely on Australian social theorist and industrial psychologist Elton Mayo's theories about the behavioural dynamics of people in large groups. According to Trahair (1984), Mayo's contested ideas—based loosely on social theories forwarded by Vilfredo Pareto and Émile Durkheim—stated that within market-industrial societies, social-relation structures (informed by scientific-management strategies) did not take into account the workers' sense of community and compassion. In the eyes of Mayo, these factors were important in workers' resistance to productivity goals set by management; they would instead seek to form their own isolated relational networks within industrial environments.

9. Conceptualised and drafted in 1785 by English social theorist Jeremy Bentham, the Panopticon is a prison architecture that allows an observer (such as a guard) to watch inmates without the surveilled subject being able to tell whether they are being observed. The goal of this device was to instil a sense of self-surveillance in the prison population, based upon the institutional threat that they were always at risk of becoming the focus of the authoritarian gaze.

10. Late in the nineteenth century, American psychologist William James and Danish physiologist Carl Lange independently proposed that emotions are triggered by physiological changes in the body, themselves brought about by experiences in the world. James explains his theory as follows: 'My thesis . . . *is that the bodily changes follow directly the* PERCEPTION *of the exciting fact, and that our feeling of the same changes as they occur* IS *the emotion*' (James 1884, 189; emphasis original). Thus, if one witnesses a dangerous animal in the wild and feels the body shake, for example, Lange and James both propose that it is this somatic reaction that triggers the emotion of fear, rather than the other way around.

11. Credited to the work of American physiologist Walter Cannon, who had been investigating the ways that bodily changes occurred in conjunction with emotions (1915), and later modified by French physician Philip Bard, the theory proposes that in an emotionally stimulating situation, emotions are initially felt, activating a part of the brain called the hypothalamus to subsequently produce physiological changes in the body.

12. Dr. Harold Burris-Meyer and Richard L. Cardinell conducted tests to ascertain how workers' fatigue patterns altered throughout the course of an average work day, charting rates of output and the times of day that workers perceived as passing more slowly or quickly. The results of the tests indicated that worker efficiency is at its highest in the morning after they start their work day and drops to a low point after midmorning before marginally picking up before lunch. The same pattern was found to occur in the afternoon, but with lower overall output than the morning. Muzak was programmed according to the dynamics of this *fatigue curve*, with the levels of musical stimulus increasing during those targeted periods of low efficiency.

Waco Sound System. *Illustration by Krystian Griffiths*

Chapter Two

Surround-Sound Manipulation at the Waco Siege

BODY OF NOISE/BODY IN NOISE

This chapter focuses on the employment by the Federal Bureau of Investigation (FBI) and the Bureau of Alcohol, Tobacco and Firearms (ATF) of sonic strategies and loudspeaker surround-sound systems in their attempts to force members of a religious sect known as the Branch Davidians out of their Mount Carmel compound in Waco, Texas, in 1993. It investigates and analyses the role played by sound in the standoff and the attempts by all parties in the conflict to amplify, territorialise and subvert the sonic domain during the fifty-one-day siege that began on 28 February and ended tragically on 19 April with the deaths of seventy-six members of the religious group.

During November 1992, ATF agents had attempted to secure warrants to search the grounds of the Mount Carmel compound belonging to the Branch Davidians, primarily on the basis of the testimonies of 'deprogrammed' and disgruntled former members of the sect. This request was initially denied on the grounds of insufficient evidence. The first complaints made against the small, fluctuating group comprised of approximately eighty men, women and children, concerned the alleged abuse of children within the community. These allegations were investigated and later dismissed by Child Protective Services of Texas on the grounds of a lack of supporting evidence. This, however, did not stop US Attorney General Janet Reno from repeatedly citing the complaints as justification for the initial ATF raid on the compound in February that was to lead to the siege and protracted standoff.

Writing on the legal implications of the Waco siege, law professor Edward Gaffney Jr. asserts that a

> major difficulty for the community, which ultimately led to its destruction, concerned the amassing of weapons at the compound. In May 1992 Daniel Weyenberg of the McLennan County Sheriff's department informed the BATF office in Austin that a United Parcel Service (UPS) agent had informed him that members of the Branch Davidian community had received shipments of firearms worth more than $10,000, inert grenade casings, and a substantial quantity of an explosive known as black powder. On June 9, a neighbor reported to the sheriff's office that he heard a noise that sounded like machine-gun fire at the Mt. Carmel compound. The sheriff also notified the BATF of this report. (Gaffney 1995, 326)

Gaffney goes on to relate how David Block, one of the most active of the apostate former Branch Davidians, had accused sect leader David Koresh of amassing and manufacturing a substantial number of illegal weapons and firearms within the Davidian compound. According to Gaffney, Block alleged that two members of the community were using a metal milling machine and a metal lathe to produce weapons, that Koresh was amassing an arsenal of weapons—including fifteen AR15s, twenty-five AK-47s and three 'streetsweepers' (twelve-gauge shotguns that rotate the magazine to position the next shot for firing)—and that Koresh posted armed guards at the compound every night (ibid.).

It is arguably this verbal testimony by Block that started the process that would eventually lead to the ATF storming the Davidians' compound.

The waveformed elements in the lead-up to the raid are of interest here, as they set a tone that was to remain constant throughout the fifty-one days that followed the first attack. As Virginia Madsen noted in her article concerning the use of light and sound as weapons at Waco, 'In many respects, the Waco 'narrative' unfolded as a series of uniquely audio events' (Madsen 2009, 90). Three types of frequency-based phenomena were predominantly at play in this narrative: (1) the spoken word in testimonial, negotiation, radio broadcast and evangelical form, (2) ambiguous noises, such as the suggested machine-gun fire in the complaint noted above by Gaffney and (3) the silence that would break out when communications between the two parties were cut, impeded or misconstrued. Together they set up a sonic trialectic between the voice, noise and silence that would be present throughout the Waco incident—a trio of acoustic states that became the foundation of the waveformed spatiality that was being fought over and territorialised by both the government and the sect.

On 11 January 1993, the ATF secured an undercover house close to the compound so they could carry out visual and sonic surveillance of the Davidians. ATF agents presented themselves at the Mount Carmel compound as students interested in purchasing equipment from the community and were subsequently invited to join a Bible class, leading to the establishment of a neighbourly relationship. After seven weeks of interaction with Koresh and his followers, on 25 February 1993 the ATF obtained two warrants: one for Koresh's arrest, the other for a search of the compound for illegal weapons. Three days later, on the morning of 28 February, the ATF sent out a mile-long convoy of heavily armed agents to execute the two warrants by means of 'dynamic entry'.

The US government had tipped off a television station about the impending strike against the Davidians in order to advertise their intention to bring law and order to the outskirts of civilisation, signalling to the wider public that the legal and sexual wilderness of the religious compound was about to be brought under control and recomposed via the legal mechanisms of the state. But in an ironic twist full of bathos, the television cameraman sent to film the martial proceedings lost his way and unknowingly asked directions from a sect member who was out running an errand. The element of surprise so important to the ATF was entirely lost when the reason for the cameraman's journey to the compound was accidentally revealed to the sect member, who then relayed the information back home to the Davidians. Realising that Koresh was cognisant of the plan to storm his ramshackle dominion, one of the undercover agents who happened to be inside the compound at the time called ATF headquarters and communicated to them that the element of surprise had been compromised and that he was not sure whether he could get out in time. This situation should have put a stop to the operation (and knowledge that the secrecy of the raid had been compromised would be consistently denied by the ATF in subsequent legal hearings and enquiries), but the decision was made to continue as planned. What ensued was a bloodbath. After one-and-a-half hours of gunfire exchange, when a cease-fire was finally agreed, both sides counted their losses: six Davidians had lost their lives, along with four members of the ATF. In the midst of the battle, Davidian Wayne Martin had called 911 and pleaded for the ATF to cease their attack because children were in the firing line. Although it was ineffective in terms of stemming the violence, this phone call was significant as it was the first voice communication between the religious sect and government agencies.

Come the evening of 28 February, the government's position had shifted, and control of the situation had been handed over to the FBI, whose immediate reaction was to try and negotiate an end to the standoff with Koresh and the Davidians. A small number of children and adults were persuaded to cross the compound's boundary and exit the sect's territory, but the majority, resisting any urges to be pacified by the 'voices of reason', chose instead to remain in place, answering the call of the divine voice from above to which Koresh would later attribute the Davidians' behaviour. Their strategy of resolution through communication having proved only marginally successful, after a few days the FBI grew frustrated, keen to conclude a situation that had now caught the attention of the local, national and international press. According to Gaffney, in order to expedite affairs, the FBI 'switched roles, from that of a "conciliatory, trust-building negotiator" to that of a "more demanding and intimidating negotiator," and began concentrating on tactical pressure, by using "all-out psychological warfare intended to stress and intimidate the Branch Davidians" . . . and by "tightening the noose" with a circle of armored vehicles' (Gaffney 1995, 328, citing Stone 1993).

This analogy of the noose is a revealing one, inasmuch as it sets up the image of the besieged stage upon which death would claim its victims. However, for a frequency-based spatial analysis, rather than the tightening noose, the notion of a sonic ring or boundary marker between the 'civilised' and the 'wilderness' is more apposite. By surrounding Mount Carmel with military vehicles, the US government reinforced a spatial distinction crucial to the Davidians themselves, for whom the compound boundaries were demarcations of body and soul, dividing lines determining whether or not one could expect salvation. For Koresh and his followers, all those who existed outside the bounds of their small community were doomed to damnation, and the initial ATF attack signalled the impending final stages of the apocalypse they believed was about to engulf the earth. As David Bromley and Edward Silver surmise in their examination of the Davidians' modus operandi before the siege, 'The group conceived of itself as literally scripting and living in the end time, a process that intensified dramatically once the confrontation with federal authorities commenced. As Koresh began to prepare for an apocalypse that he increasingly thought might occur in America rather than Israel, the group began adopting survivalist tactics such as stockpiling large amounts of dried food and MREs (meals ready to eat) used by the military, weapons and ammunition, and a large storage of propane gas' (Bromley and Silver 1995, 61).

With the media unable to conduct interviews or to document the siege from the perspective of those inside the compound, the Davidians found themselves silenced and unable to relate their version of the events that had taken place in the preceding weeks. As noted by James Richardson, the FBI tightly controlled all visual access to the site: 'When enterprising reporters did seek to get closer than the three-mile limit, they were treated harshly. A number of photographers, tired of the "lens wars" that had developed as media outlets sent stronger and stronger lenses to Mt. Carmel, violated distance limitations imposed by the FBI. . . . They were summarily arrested, thrown to the ground, handcuffed, and taken away to jail' (Richardson 1995, 165).

The FBI would have dearly liked to do the same with Koresh, especially after the interviews he unexpectedly gave on the first day of the siege. Isolated and alienated, Koresh had been forced to amplify his voice by telephone, speaking over a live connection to the KRLD radio station in Dallas and via CNN on cable TV. These interviews with Koresh would be the first of 'hundreds of hours of explanations, based on his understandings of the biblical apocalyptic significance of the situation in which he found himself' (Tabor 1995, 263–64).

On the day of the initial raid, negotiations between the Davidians and the FBI had also been conducted over the telephone, a channel that would continue to see heavy traffic over the coming weeks: 'According to FBI records, during the fifty-one-day period, negotiators spoke with fifty-four individuals inside Mt. Carmel for a total of 215 hours. There were 459 conversations with Steve Schneider, which consumed ninety-six hours. Koresh spoke with authorities 117 times—a total of sixty hours' (ibid., 265).

It became evident that the most important form of communication relay between all concerned would be sonic, taking not only the form of telephone calls but also audiotapes, radio programmes and, later, barrages of music over loudspeakers. As the ongoing sense of alienation deepened, Koresh requested that tapes containing his spoken-word monologues be aired over the radio so that his understanding of the Bible could be transmitted to an audience outside of the FBI, whom he believed had little comprehension of his religious beliefs. In fact, the FBI was quite willing to mediate the request, allowing that 'some of Koresh's "ramblings" be played on a radio station, as Koresh had asked, in order to try to gain his surrender. FBI officials became upset when Koresh called CNN directly at one point, and stopped the activity immediately by cutting all phone lines except the one they wanted kept open' (Richardson 1995, 165).

As this example suggests, accessing, expanding and controlling the channels of sonic latitude became an increasingly important and fractious business as the siege went on, while the sonic longitude of which Koresh perceived himself to be at the somatic end—the communication channel that, he insisted, patched him into the voice of his God—fell on deaf ears within the FBI, who had little time or space for the notion that such a channel might exist. By 21 March the communicative amplitude of those within the compound became a physical and psychological force to be reckoned with, and the FBI realised that the antenna-body of Koresh, as he spoke with the media, extended beyond the physical boundary they were patrolling with armoured vehicles. As the results of its failure to understand this dynamic became clear, the FBI changed tactics, effectively remapping the acoustic dynamics of the standoff by severing the phone lines and opening up a new, one-way channel by surrounding the Davidian compound with speakers and barraging those inside with noise, music and sound effects. Bright lights were also shone on and into the compound at night, producing a blitzed sense of never-ending days. Negotiation and exchange were no longer on offer in this radically altered, bleached-out and waveformed landscape.

Those within the Branch Davidian complex were now obliged to continuously listen, with no opportunity for recourse or reply. The FBI meanwhile amplified its new mode of territorialisation. This was a sonic technique that was invested in the mastering of volume, repetition and duration. Its aim meanwhile was clear and simple—to render passive those compounded targets deemed to be a lethal threat. 'From the federal agents' perspective', write Anson Shupe and Jeffrey Hadden,

> the early days of the siege confirmed their understanding of David Koresh as devious, manipulative, unreliable, and extremely dangerous. In the course of the fifty-one-day siege, agents pursued an unusual number of psychological-warfare measures. The FBI tactics included the use of loudspeakers to bombard the Davidians with propaganda and harassment. Presumably the agents held out some hope that they would overcome the grip that Koresh held over his followers. At all hours of the night and day, the loudspeakers belched forth such curious content as audiotapes of rabbits being killed, chanting Tibetan monks, and Nancy Sinatra singing 'These Boots Were Made for Walking'. (Shupe and Hadden 1995, 189)

Disorienting the Branch Davidians, silencing them and depriving them of sleep—this strategy of sonic attack—documented by Ronald Cole in his White House paper *Operation Just Cause* (Cole 1995)—was only discontinued when the Dalai Lama intervened and demanded that the employment of sacred Buddhist music for martial purposes cease.

One of the most intriguing and pertinent lines of enquiry about the use of such a strategy in this ideological conflict opens up with the seemingly perverse question: Who deejayed Waco? This apparently glib query assumes a deeper level of resonance when we ask whether there was any tangible strategy exercised here by a government agency proposing itself to be the legal, moral and fundamentally rational party confronting the supposed irrationality of 'religious fundamentalists'. Whichever think tank or individual composed the playlist for the sonic content broadcast at Waco made choices that veered between the camp, the profane and the bizarre. As a result, the 'standoff set list' reads more like an art student's sound-art installation than a military strategy—but then maybe the two are not as different as we might like to think.

CHANNELLING VIOLENCE

In our 2010 article 'The Art of Conservative *Détournement*', art historian Andrew Hennlich and I argue that cultural forms of expression are increasingly being utilised for military purposes:

> The Operational Theory Research Institute, an Israeli Defence Force 'think tank' directed by Shimon Naveh, turned to the philosophy of Guy Debord, Gilles Deleuze and Félix Guattari, the architectural work of John Forester, Bernard Tschumi and Clifford Geertz and the 'Anarchitectural' site-specific urban interventions of Gordon Matta-Clark to facilitate the respatialisation of contemporary military theory and strategy. Upon further inspection, we are able to map a wider system of cultural and ideological assimilation through a range of military organisations that employ theories and works from the traditionally perceived humanitarian disciplines of music, architecture, art and philosophy. Examples include the US military's use of music for 'battlefield preparation' as well as for torture in Guantánamo Bay and Abu Ghraib, and Canadian military training centres such as 'Pretendahar' in Toronto which prepare soldiers for combat in the Middle East, referencing 1990s installation-art practices. These examples give adage to the notion that this is not military 'business as usual' but rather the martialing of the business of culture. (Heys and Hennlich 2010, 61)

In the case of Waco, the FBI playlist incontrovertibly expresses the nonlinear context from which its selections were drawn—namely, a culture in the throes of postmodernism, a global phenomenon described by Fredric Jameson as 'the internal and superstructural expression of a whole new wave of American military and economic domination throughout the world'. Jameson goes on to add that 'in this sense, as throughout class history, the underside of culture is blood, torture, death, and horror' (Jameson 1991, 57). As a strategy, the Waco playlist harkened back to an earlier psyop[1], 'Nifty Package', which had taken place almost four years earlier in December 1989. In this case the US military had surrounded the papal nunciature—the Vatican diplomatic mission in Panama City—with speakers because General Manuel Noriega had been granted refuge within it after a $1 million reward had been offered for his capture. Day and night the speakers played tracks by artists such as Van Halen and Judas Priest at high volume precisely because Noriega was known to loathe rock music. On 3 January 1990, the general gave himself up. Whether it was the barrage of music that forced him to surrender is still open to debate, but in any case, here the surround-sound strategy of psychological harassment had been deployed in an international political context. In marked contrast, the Waco siege represented the first time that such a strategy had been employed by the US military on home soil, against its own citizens. This act of 'turning on one's own', of using strategies previously deployed against those considered to be culturally and politically external 'enemies', reveals much about the ways in which techniques that are tried and tested overseas by militaries are subsequently relayed and modified for utilisation by internal martial and enforcement agencies against their own citizens. Paul Virilio traces this blurring of the boundaries between civilian and militarised psychologies and spatialities back to

> the new 'secret police', which Balzac considers the most important social revolution of his time—the moment when, after the long period of ostensible and bloody repression exerted against the civilian populations by the Revolution's 'army of the interior', military violence stops being necessarily visible only from afar, by the soldier's uniform, and comes to rest on refined systems of surveillance and denunciation. These first attempts at penetration, clandestine 'invasion' of the social corpus, had . . . a specific aim: exploitation by the armed forces of the nation's potential (its industrial, economic, demographic, cultural, scientific, political, and moral capabilities). Since then, social penetration has been linked to the dizzying evolution of military penetration techniques; each vehicular advance erases a distinction between the army and civilisation. (Virilio 2006, 124–25)

In the case of the Waco siege (and, more generally, any internal military actions against one's own citizens), the boundaries of sonic conflict are inverted, in terms of both the targets' identity and the type of weaponry used against them. It could also be surmised that, by using music as a territorial and psychological weapon at Waco, the government was in fact engaging in conflicts in order to assimilate and regulate two societal elements—religious fundamentalists and musical cultures—both deemed to have the capacity to provoke and threaten constitutional authority.

This suspicion—that what is at stake is the government's pacification of cultural products believed to hold subsersive or resistant potential—is corroborated by the fact that this is certainly not the first time art, architecture, music and philosophy have been co-opted and utilised by military organisations. Such usages have a history that can be traced back to at least the early twentieth century. For instance, Slavoj Žižek notes that in 1938 the Franco regime used architectural practices informed by Surrealism to construct a 'series of secret cells and torture centers . . . in Barcelona' (Žižek 2006, 3). It is important to take note of such early precedents, yet they fall well short of the systematic implementation of military strategies based on assimilated cultural ideologies and practices that we see in more recent uses of music as torture, the appropriation of installation art practices for training centres and the use of avant-garde philosophies as manuals for martial manoeuvres. In these contemporary innovations we see an inversion of enemy territory as the military travels inside and mines its own civilian network in order to negate threat, camouflage landscapes of resistance and draw up new, inverted cartographies of culture.

After the sound and light attacks that were mapped onto the compound failed to dislodge the entrenched Davidians, the FBI professed to have tried to 'move heaven and earth' to resolve the standoff. According to the FBI, the fact that the Davidians 'held out against psychological-warfare tactics demonstrated that [they] were dogmatic, determined, and unrelentingly devoted to their leader', the logical conclusion being that, 'for their own safety and well-being, they had to be rousted out by any means necessary' (Shupe and Hadden 1995, 189). As noted by Madsen, this pathological determination of the Davidians was matched by those surrounding the compound and was consumed by a global audience that witnessed the drama unfolding:

> For fifty-one days the eyes of millions of spectators were upon Waco. This was a site subject to negative development before the media's 'absent' eye. For fifty-one days, the international news media waited. For the TV cameras—hungry for lights, action, exposure—there was little to develop, little to be seen on the outside except the boarded-up white building on a treeless Texas plain. ... Images, sounds, and stories could repeat themselves, creating feedback loops and dangerous sympathetic vibrations. As the days dragged on, tensions built, not only were the officers growing weary, so too were audiences and media. In their paranoia, the Branch Davidians were not to know how time could catch up with them—that 'the End', their Apocalypse, could come at such speed, and so soon. They had dug in for the long duration, as if duration still counted. (Madsen 2009, 98)

As we know, duration ceased to be an issue for the Davidians on 19 April 1993, when the government ordered tanks to breach the walls of the compound and to disperse canisters of CS (tear) gas throughout the buildings in an effort to force the desperate inhabitants out. Even at this point, when the conflict became face-to-face, sonic strategies came to the fore, as the FBI covertly placed small powerful microphones through the wounds of the punctured building, eavesdropping on the Davidians as they prepared for a death that they could not have imagined yet that must in some way have entreated their apocalyptic expectations. The ensuing violence and the manner in which it has been culturally memorialised, analysed and rendered as a cultural product was aptly preempted by Jacques Attali when he declared that 'eavesdropping, censorship, recording, and surveillance are weapons of power. The technology of listening in on, ordering, transmitting, and recording noise is at the heart of this apparatus. The symbolism of the Frozen Words, of the Tables of the Law, of recorded noise and eavesdropping—these are the dreams of political scientists and the fantasies of men in power: to listen to, to memorise—this is the ability to interpret and control history, to manipulate the culture of a people, to channel its violence and hopes' (Attali 1985, 7).

The blistering noise caused by the all-consuming fire that would break out and kill the majority of the remaining inhabitants of the Mount Carmel compound was to become the final aural testimony of the siege. All of the material and ocular evidence of lives once lived in the compound were converted into searing frequencies, the waveforms of the flames becoming the ultimate sonic signature of a catastrophe that had been essentially orchestrated, recorded and played back within the acoustic domain. And as for those in power, alluded to by Attali more than thirty years ago, they have certainly not been able to control history in the way they might have wished, there being a surfeit of films, miniseries and books contradicting the US government's version of events at Waco[2]. But in terms of channelling violence, they have found new ways of orchestrating this, and as we shall see in later chapters, asymmetric modes of waveformed violence seem to be at the forefront of their fantasies.

A PSYCHOLOGICAL SYMPHONY OF DISASTER

The Waco siege presents us with a high-profile event in which music and noise were employed for the purposes of psychological manipulation in order to regain command of a situation deemed to be out of control. The fortified compound caught in the grip of confusion and conflict is the antithesis of the managed social laboratory of the Fordist factory discussed in chapter 1, a controlled and ordered test site for psychologists' and sociologists' experimentation and recording. The set of legal, architectural and humanitarian parameters in evidence at Mount Carmel differed in almost every way from those of the factory workplace, the only ostensible link between the two being the utilisation of music to try to govern a targeted social group's psychological and physiological actions. In order to induce acceptably pliant behaviour in the Davidians (whose passive, anxious dispositions would have been more likely to result in their surrender), the FBI treated the architecture of the compound as a kind of magnified version of the operant conditioning chamber, or Skinner box, invented by American psychologist Burrhus Frederic Skinner in the 1930s. A piece of laboratory apparatus used to study animal behaviour, the Skinner box is designed to enable the controlled administering of stimuli and measurement of a subject's response to them, the central proposition of Skinner's behaviourist theory being that when a certain response to a stimulus is rewarded, the pattern is reinforced and the subject is subsequently conditioned to respond in the same way again in the future. As outlined in his text *Radical Behaviorism* (1953a), this element of reinforcement is a key factor in attaining the desired responses that are aimed for in any given experiment that employs the principle.

Early exchanges between the FBI and Koresh implied such a system of reward and reinforcement, as the FBI agreed, for example, to have Koresh's taped monologues aired on the radio with a view to securing the sect's eventual capitulation to government demands. But when Koresh reneged on the agreement struck between the two parties, the government quickly changed their psychological and territorial position within the waveformed terrain and, instead of offering rewards to lure the Davidians out, turned to different, forcible means of behavioural influence that amounted to sonic punishment. Upon this strategic change, an audible schism opened up in relation to the way that the FBI's actions referenced Skinner's theory: Instead of stimuli being presented in small amounts so that responses could be reinforced or 'shaped', sonic information was constantly amplified in repetitive sequences over an extended period of time. Such an overt shift in the psychological dynamics of the siege—from the FBI's measured reinforcement of its position (plotted through numerous phone calls) to repetitive punishment—made explicit how easily the sonic sphere could be reorganised and weaponised; it was this recomposition of the movements and interactions of the siege that resulted in their coming to symbolise musical notes in an ensuing symphony of disaster.

From the boxes carried by conveyor belts in chapter 1 to the Skinner box, from the social grouping of the workforce to that of the sect and from the architectural context of the factory to that of the compound, the waveformed event addressed in this chapter concerns a smaller and more closely knit social body that existed in an extended domestic setting. On a daily basis, the sect functioned on more physically, psychologically and emotionally intimate levels than did the members of the workforce in the Fordist factory. Perhaps because of this tighter interpersonal bonding, the frequency-based techniques and strategies used against the Davidians in Waco can be perceived as more intense, and more heavily amplified, than those put to work in the factory.

Koresh understood only too well the power of music to influence and entrain people. He had invested heavily in his persona as an outlaw Christian rock figure to channel his charisma and to make it pay off in the currency of his followers' loyalty. Taking on the role of a rebellious Christian rock star who had little time or space for the laws and norms of mainstream society enabled him to further psychologically influence those who had already been seduced into the Branch Davidian community and to captivate and exert an allure over those he met outside of the compound. Rock 'n' roll may never die, but it must always be intimately tied to destruction—of the self, of the other and of anything else that happens to get in the way—and it is difficult to think of many who embody this sonic mythology as fully as David Koresh. As Tipper Gore might have said in her glory days, *Christian rock is dangerous for your child's health.*

THE FIFTY-ONE-DAY DIARY OF A FAILED ROCK STAR

Having left Hollywood after a failed attempt to find fame as a musician, Koresh joined the Branch Davidians and, having gained leadership of the group, proceeded like a latter-day Pied Piper to seduce men, women and children to leave their families, jobs and homes (to which the majority would never return) and join him in the Mount Carmel compound. The figure of the Pied Piper who uses music to seduce people away from the safety of their homes and loved ones is a common reference in contemporary Western culture when discussing controversial figures perceived to rail against mainstream values—from famous convicted criminals such as Charles Manson and schlock horror rock stars such as Marilyn Manson to more obscure serial killers such as Charles Schmid, known as the 'Pied Piper of Tucson' (see Gilmore 1995). The suggestion here is that inexplicable powers may drive people to carry out unspeakable acts and that such powers can be utilised to secure the cooperation and obedience of others. The belief that such powers can be harnessed by music effectively helps us rationalise extreme forms of violence that we consider to be inhuman, giving us a language with which we can speak of the ineffable.

As Koresh increasingly came into conflict with mainstream North American values—by practicing polygamy and taking wives as young as twelve years old (with their parents' consent)—the sect became further estranged and alienated from the conventions of the dominant society from which its members had been drawn. Identified as a 'cult' by those fearful and suspicious of the ethical independence they were staking out for themselves (suspicions fed by the rumours that rape and child abuse were routine occurrences within the compound, certain of which proved to be largely unsubstantiated), the Davidians were pushed over the threshold of what was considered acceptable by mainstream society.

But the sect itself already conspicuously sought to be outside of the legal, moral and earthly mores that preside within the semipermeable boundary marker of 'civilisation'. Often within this dynamic, mainstream society reinforces the expulsion of a group by identifying it as a 'cult', a designation that has proven throughout the past century to have severe repercussions for its members. The branding of the Davidians as a cult was pivotal to the ways in which the Waco standoff was perceived and mediatised and the reactions it elicited, as confirmed by Shupe and Hadden: 'This image of cults is firmly anchored in American history but has been recently reinforced by the plethora of new religions over the past two decades. Virtually everyone knows about the mass suicide of hundreds of followers of Jim Jones in Guyana. But many other groups, such as the Unification Church (Moonies), Hare Krishnas, the Children of God, and Scientology, have been the subjects of high-profile and highly negative publicity. Hence, the mere labeling of a group as a cult conjures up the most scandalous of images' (189).

As John Hall argues, 'Anticult usage identifies "cults" as subversive both of families and individuals and of the core values of society as a whole' (1995, 209). Given that their value system was indeed intrinsically distanced from such core values, the apocalyptic sect had concluded that they needed to cut any remaining bonds tying them to mainstream society in order to achieve micronation-style autonomy[3]. To the extent that groups such as the Davidians realise that their behavioural codes, sexual practices, familial formations and shared economic practices will not be tolerated or deemed acceptable by mainstream culture, there is all the more incentive to work entirely outside of it. Marginalisation allows a group to formulate new ways of relating to the entire social spectrum of actions and exchanges, as well as redefining the inevitable divisions, hierarchies and ways of forming leadership that are endemic to any group. When mainstream culture identifies such a transgression and cements it by labelling a group as being outside of their systems of socialisation, justice and economics, a social contract grounded in banishment is publicly engineered.

The sociopolitical compact between mainstream and 'outsiders'[4] is made at the point of rupture, when those who have been cast out not only accept their position and context but also audibly and visibly celebrate their deviation and separation from mainstream value systems. James Lewis notes how 'deviants come to symbolise or personify changes, threats to the status quo, where "the community is confronted by a significant relocation of (moral) boundaries"' (Lewis 1995, 102, citing Erikson 1966, 68). For society at large, when those who have become 'outsiders' or 'deviants' can be identified as such, they can then be labelled as 'enemies' in the final stage of a social diagnosis that creates and formats 'opponents'. Lewis describes the dynamics of this phenomenon: 'One of the more widely accepted dictums of sociology is that societies need enemies, particularly societies that are going through a disturbing period of change. External threats provide motivation for people to overcome internal divisiveness in order to work together as a unit. Conversely, "the unity of the group is often lost when it has no longer any opponent". Having an enemy one can portray as evil and perverse may provide social solidarity and support for the normative values and institutions of one's society' (Lewis 1995, 102).

Mainstream culture requires social peripheries that function by delineating, enforcing and reifying its social bonds of morality, sexuality and friendship. By pushing individuals and groups outside of this inclusive boundary, it creates otherness and the 'outsiderness' explored by Rhys Williams:

> In his important book *Religious Outsiders and the Making of Americans*, Laurence Moore argues that 'outsiderness' is more than just a common feature of religious minorities in United States history. Rather, he maintains that outsiderness has been a necessary aspect of the struggle to establish new religions in this country. Moore argues that outsiderness has often been promoted by the groups themselves, eager to maintain a distinct identity even as they try to find a niche in American society. Moreover, this 'self-stigmatization' has been so common that it is part of what it means to be an American; the often-uneasy balance between group distinctiveness and social accommodation defines American religion. (Williams 1995, 301)

Further delineating the development of the identity of an alienated sect perceived to be threatening to mainstream culture, Dean M. Kelley affirms that 'the communal group may develop distinctive modes of dress, a coded in-house language, unconventional domestic arrangements, and other peculiar traits, and the result between them and others can be mutual dislike, suspicion, fear, and stereotyping. This growing antipathy serves to increase the religious group's alienation and distrust to a degree approaching paranoia' (Kelley 1995, 363).

The group psychology of the Davidians as outsiders was to play an important role in the 1993 siege. By constructing their own living quarters, value systems and waveformed environment of collective music making, their internal social bonds were strengthened to a degree that governmental agencies considered 'unnatural' and unlawful, calling for armed and frequency-based intervention.

With all channels of communication to the outside world cut off (other than those opened by the FBI), the psychological dynamic between the compound's inhabitants became increasingly important to the authorities as they attempted to depict Koresh as losing control over his followers. To further this aim, excessive physical force deployed against the Davidians was transferred to the sonic domain via the speaker systems placed around them. Sound from the sect was no longer understood to be transmitting into the world beyond the speakers; it was instead trapped in a feedback loop. Feedback of the linguistic variety, meanwhile, was cut off, gated at the boundary where all utterances, vibrations and signals were monitored by government operatives with their surround-sound systems, microphones and technical expertise.

Invoking the ideas of Marc Galanter, Thomas Robbins and Dick Anthony examine the functions of various types of feedback, in particular feedback that occurs in an isolated social group where the psychological demarcation of boundaries oscillates dramatically when excessively questioned:

> A cult or charismatic group operates as a social system. It has a primary task or basic transformation function, which entails transforming input from the environment into a form that meets system needs, that is, converting and socializing recruits. A monitoring function monitors, regulates, and coordinates the action of component parts of the system. A feedback function enables the system to obtain information on how effectively it is carrying out its primary task. Negative feedback is vital to long-term self-regulation of the system but also poses a short-term threat to the system by undermining participants' moral and sometimes challenging group beliefs. The suppression of negative feedback is a constant temptation, which, however, may ultimately impair the system. Finally, charismatic social systems have a dimension of boundary control, protecting the systems from external threats. Most outbreaks of violence associated with religious movements entail escalating boundary tension. (Robbins and Anthony 1995, 249–50)

VIRASONIC ATTACK

Aiming to compound the levels of anxiety in the Davidians' environment, the FBI programmed its sonic attack from the compound's boundary, attempting to break down the psychological bonds that were enforced by collective singing, prayer and music making, as well as by those (micro)sounds generated by members' daily communal activities. The governmental walls of sound produced at Waco amplified mainstream society's desire to estrange themselves from the Davidians' belief systems and from Koresh's reading of and subsequent communication of the Seven Seals. Although referring more broadly to psychological shifts brought about by Occidental culture's amplification of sound, R. Murray Schafer's statement about music's capacity to alienate is germane here: 'Walls used to exist to isolate sounds. Today sound walls exist to isolate. In the same way the intense amplification of popular music does not stimulate sociability so much as it expresses the desire to experience individuation . . . aloneness . . . [and] disengagement' (Schafer 1993, 96).

The play between naturally occurring noise and musically constructed sound would have spatially and temporally anchored the Davidians, reaffirming their identity through ritualistic practices such as the regular prayer meetings that were held up to three or four times a day. Inducing sonic chaos so as to break down this religiously timetabled existence served not only to separate the members from each other but also to sever the c(h)ords that attached the members to the waveformed patterns of their natural environment. The psychological games played by both the American governmental forces and the ostracised Davidian sect during the standoff played out like grotesque theatre, albeit with a networked audience of millions. The stage was set up as a contemporary Western shoot-out with two identifiable foes facing each other, both on the verge of extreme violence, the anti-Apollonian soundtrack and extremely harsh lighting provided by the FBI, and the press (as live audience) clamouring for exclusive footage. Sociologist James Richardson indicates the seeming inevitability of both the outcome and the mediatised evolution of events when he declares that 'the evening news had all the elements to guarantee high ratings—religion, sex, guns, child abuse, and other violence, in tandem with the intrigue and complex plotting by both sides. The drama seemed to imitate a Greek tragedy, moving inexorably toward its predictable climax, and we know that Greek tragedies always involve predetermined sacrifices. That tragic sense of the "meaning" of the Waco tragedy was borne out on April 19 when authorities did what Koresh had been predicting all along to his followers' (Richardson 1995, 164).

Commentaries on the ways in which music was used at Waco are routinely predicated solely on the deployment of frequencies against the Davidian sect. Yet there are also questions concerning how the repetitive musical bombardments psychologically affected agents patrolling the perimeter of the Mount Carmel compound. As we know, it is difficult to shield humans from frequencies of any kind, especially those in the lower audible to subsonic range. As well as the volume levels (which would have been remedied by wearing ear protection), there is the issue of the repetition of the sonic content and its corollary effect on the cognitive and emotional behaviours of those orchestrating the sonic payload of Nancy Sinatra and dentist's drills. The analogy most useful here is that of 'friendly fire'—in the case of Waco, perhaps we should consider the consequences of what could be called 'friendly frequencies'.

Radically remapping the acoustic terrain of the siege, the mix of noise and music that was transmitted to cause anxiety among the Davidians would also have created distance between the FBI agents, as communications must have become more difficult in the wake of such volume levels. Within periods of armed conflict, such as the siege, camaraderie is often composed through verbal communication, so when this is replaced by repetitious noise, a perverse silence is produced in its place. This seemingly contradictory sonic duality renders a new synthesised version of Virilio's theatre of operations in which distance is also created internally between allies, prompting us to reevaluate his contention that 'the battlefield is the place where social intercourse breaks off, where political rapprochement fails, making way for the inculcation of terror. The panoply of acts of war thus always tend to be organised at a distance, or rather, to organise distances. Orders, in fact speech of any kind, are transmitted by long-range instruments which, in any case, are often inaudible among combatants' screams, the clash of arms, and, later, the various explosions and detonations' (Virilio 1994, 6).

The only screams heard during the first fifty days of the siege were those of dying rabbits, played at high amplification levels repeatedly, day and night. The question of whether such transmissions had detrimental effects on FBI personnel has not been answered, but the fact that these agents held dominion over the sound systems would have made it easier to endure, as they had control over the volume, duration, repetition and content throughout the operation. As shall be explored further in chapter 3, in the case of the US military guards in Guantánamo Bay who similarly organise and witness sonic offensives, the agency held by the perpetrators is crucial to the reception and endurance of extreme sensorial conditions such as those generated at Waco.

One of the key components of anxiety- and fear-based psychological strategies such as the FBI's sonic assault is the creation of apparent chaos. When a group's routines and sonic rituals are not only disrupted but also made entirely unviable, then bonds, trust and interaction have to be remodelled around other activities and nonverbal communications rechannelled to provide belief-affirming exploits for group members. Madsen appraises the capacity of sonic content used at Waco not only to challenge group belief but also to potentially cultivate alliances. She writes, 'From a physical and psychological "point of listening" . . . the many unorthodox sound recordings used against the Branch Davidians could reveal a highly affective power, with the potential to induce disintegration or dangerous "sympathetic vibration"' (Madsen 2009, 93). When the aural conditions for constructing socially bonding practices are drowned out, the natural harmony and rhythm of a group's social life is disrupted and called into question. Knowing what is going to come next makes it easier to prepare oneself and to formulate courses of action ahead of time in any given situation. When one is forced to constantly react to newly modified sonic contexts produced by an array of speakers, a cognitive disarray is psychologically (de)composed, as a sonic entropy put to work within the habitual system of 'making sense' distorts perceptions of agency.

INVERTING WILDERNESS

What happens, then, when speaker systems are relocated from a building's interior (as in the factories studied in chapter 1) to its exterior to be used for strategic purposes, as they were during the Waco siege? There is an obvious transmutation of affect when speaker systems are employed outside of a defended site so that frequencies—and the hardware that amplifies and delivers them—merit the dubious honour of being labelled 'weapons', an epithet hardly applicable to the Muzak-filled speakers of the Fordist factories. The external use of waveforms against an architecturally contained target calls into question the capacity of those defending the building to prevent adversaries from breaching its physical, social and psychological borders. Since frequencies do not abide by the parameters of physical movement dictated by material constructions, the use of sound is a perfect way to psychologically disrupt the solidity and safety of physical, tangible and therefore defensible architectures—although it can equally be used to support them. Waveforms transgress physical structures, yet they are absolutely connected to the physical. Schafer understands well the volatile nature of their power, as shown by his suggestion that it is more a question of channelling than of amplification: 'The important thing to realize is this: to have the Sacred Noise is not merely to make the biggest noise; rather it is a matter of having the authority to make it without censure. Wherever Noise is granted immunity from human intervention, there will be found a seat of power' (Schafer 1993, 76).

The seats of power at Mount Carmel—administratively located in Washington, D.C.—introduced both light and sound-based strategies to the siege, rendering it metaphorically akin to an (un)social sciences experiment. Surrounded by sound systems, recording devices and media technologies, the Branch Davidians were treated like insects to be sonically dissected; with macrosound instead of microscopes and tweeters rather than tweezers, the FBI pinpointed its target as a scientist might skewer a bug, but with sonic apparatuses instead of surgical steel: 'One of the primary objects of discipline is to fix; it is an antinomadic technique' (Foucault 1975, 218). The sonic discipline of the compound/laboratory at Waco was certainly meant to fix the Davidians in the temporal and spatial nexus of the dead end. And to apply the necessary psychological pressure that would make them also understand this to be the outcome of the holding pattern they were in, the FBI endeavoured to formulate a frequency-based wilderness inside the loop of the surrounding speaker system—a recorded and recordable sonic chaos that would serve the government cause by providing a waveformed space in which to experiment and document the sect.

To the inhabitants of the compound, the hinterlands that lay beyond the limits of Mount Carmel would have been seen, to various degrees, as a symbolic wilderness. To some, the edge of the compound would have been perceived as the boundary where the Davidians ceased to organise space and where nature subsequently took control—a geographical expression of the point at which 'civilised' human spatiality meets nature in the form of the 'wild'. To others, the boundaries of the compound would have signified the circumference of the sect's economic, legal, sexual and social law, conferring the role of the moral wilderness upon the mainstream society that lay beyond. The wilderness that is of most interest here, however, is of the frequency-based variety—the point at which the sonic terrain becomes unknowable, threatening and alienating. By cutting off access to the quotidian sounds of the surrounding environment, of bird calls, animal sounds, traffic reverberations—so that the previous sonic coordinates responsible for guiding one's journey through time and space were replaced by preprogrammed abstract noises and music—the FBI inverted the Davidians' perception of wilderness. They remapped the compound as existing at the centre of a waveformed wasteland, a spatiality in which the sounds that had helped orient and connect those inside were silenced and where new aural systems of isolation threatened to confuse and bewilder them.

Delineating new sonic territories via noise or music can be understood as a way of circumscribing the extent of one's power while also renegotiating the extent of the aural contract with one's natural environment. For Attali, noise belongs naturally within the realm of functionality and potency, as made clear by his assertion that 'among birds a tool for marking territorial boundaries, noise is inscribed from the start within the panoply of power. Equivalent to the articulation of space, it indicates the limits of a territory and the way to make oneself heard within it, how to survive by drawing one's sustenance from it' (Attali 1985, 6). Rather than articulating the will to survive, the marking of space via noise at Waco presaged the Davidians' inevitable demise. Unable to find a way to negotiate the wilderness that had been imposed upon them, their swan song was to be dismally serenaded by the sirens of fire trucks rather than by the trumpets of angels.

Demarcating the channel between wilderness and salvation, the ring of speakers around the compound also played on ancient techniques of mapping the environment via sonic cartography. Anthony Storr describes one such venerable technique practiced in Australia, referencing nomadic writer Bruce Chatwin's firsthand research into such phenomena: 'Bruce Chatwin, in his fascinating book *The Songlines*, demonstrates how songs served to divide up the land, and constituted title-deeds to territory. . . . Aboriginals used songs in the same way as birds to affirm territorial boundaries. Each individual inherited some verses of the Ancestor's song, which also determined the limits of a particular area. The contour of the melody of the song described the contour of the land with which it was associated. As Chatwin's informant told him: "Music is a memory bank for finding one's way about the world"' (Storr 1992, 19–20).

The contours of the government's melodies did not so much define the land and one's connection to it as signify the end of its comprehension. Hoping to sever the c(h)ords that tied the Davidians to their environment, the FBI ushered in the formulation of new disruptive and intimidating memories, placing their targets in a state of aural dislocation as they lost their everyday acoustic markers.

The condition of displacement proposed here is important in Christian history, a fact that cannot go unmentioned given the religious affiliations of the sect and its members' decision to place themselves in a lacuna between 'civilisation' and 'wilderness'. J. M. Coetzee examines the conflicting meanings inherent within the latter term, helping form a better understanding of why the FBI's decision to alienate the sect was not a prudent or effective strategy:

> The wilds, 'the wilderness' are resonant words in the Judaeo Christian tradition. In one sense, the wilderness is a world where the law of nature reigns, a world over which the first act of culture, Adam's act of naming, has not been performed. The origins of this conception of the wilderness lie in pre-Israelite demonology, where the wilderness (including the ocean) was a realm over which God's sway did not extend. But a second sense of the wilderness grew up in Judaeo-Christian theology: the wilderness as a place of safe retreat into contemplation and purification, a place where the true ground of one's being could be rediscovered, even as a place as yet incorrupt in a fallen world. (1988, 49)

For the Davidians, the grounds of Mount Carmel were a protected sanctuary from a world they believed was about to fall into the throes of Armageddon. A sect structured around its own alienation from the rest of society, the fabric of their communal ties cut from the cloth of necessitated estrangement, the Davidians had already imposed upon themselves a retreat to a position 'outside', a religiously denominated territory on the cusp of the wilderness, the social and the living.

THE ACOUSTIC ONTOLOGY OF AN OUTSIDER

Music from the outside, or 'outsider music', designates frequency-based expressions conceived of as coming from the periphery of a Western culture that has eventually (and inevitably) tired of what is known and subsequently listens for the echoes of those who live and work on the social margins or beyond, feeding an ever-growing hunger to locate, assimilate and take advantage of the most esoteric forms of sonic expression. To secular Westerners, religious music such as Christian rap and rock belongs firmly on the edges of culture, yet it is the church that has a history of marginalising and demonising waveforms, especially when they are organised into musical compositions. From the 1960s to the 1980s, controversy, predominantly stoked by religious groups, regarding backwards masked messages in commercially released music was regularly in the news. Historically, however, it is the tritone—a musical interval that spans three whole tones—that initially elicited the wrath of the church. Around the turn of the eighteenth century the tritone, owing to its musical dissonance and alleged provocation of sexual feelings in the listener, earned itself the enviable name of *diabolus in musica*, or 'the devil in music' (F. J. Smith 1979). Given this 'outsider' history, the tritone was always going to become musically popular with those who deem themselves to be socially alienated.

At the other end of Christianity's sonic spectrum is the choral (and the chorus)—those harmonious renderings in music that facilitate the bringing together of people, united in a feeling of empathic adulation towards that which is understood to be greater. The church was, after all, the first institution to recognise that 'music has the effect of intensifying or underlining the emotion which a particular event calls forth, by simultaneously coordinating the emotions of a group of people' (Storr 1992, 24). The verifiable existence of musical genres, movements or tones that are linked with 'good' or 'evil' is not what is at stake here. Of more interest is the premise that music can also become a vital factor in the process of marginalising an adversary. The identification of individuals or groups as other, as peripheral or as outsiders is manifested via signifying traits that can be openly perceived by others so that alliances can be formed and joined. Musical affiliation and adoration marks, fractures and reconstitutes social factions like no other cultural form of expression. The power of music to demarcate sociopolitical, economic, psychosexual, physiological and geographic territory is therefore considerable. David Koresh knew this very well while adopting his rock-star persona and leading guitar-fuelled Christian sing-alongs to unite his followers. In doing so, not only did Koresh attempt to draw his followers further outside of the culture from which they had come, he also attempted to induce in them a sense of being outside of their bodies, thus bringing them closer to the spiritual realm to which he supposedly held the key.

Julian Henriques explores the ability of music to take one outside of one's normative somatic, social and sexual context in the case of Jamaican dancehall sound-system culture, yet the following quotation could be speculatively describing a religious sect's nightly ritual as they sonically embrace and celebrate the transgressive agency inherent to their position on the margins: 'Sonic dominance transduces the crowd across several thresholds over the passage of the session through until dawn. It generates a very special type of environment and experience—a place between places and a time out of time. Anthropologist Victor Turner describes these as "liminal" states or places. Outside normal society, these are thresholds where transition, transfigurations and rites of passage occur. In such liminal states communication often takes place at a sublime or heightened level' (Henriques 2003, 469).

Since both Koresh and the FBI were interested, for opposite reasons, in drawing the compound's inhabitants into a sonically generated liminality, the sonic domain became the battleground where the devil's dissonance and Christian harmonies mixed into one another on the periphery—an aural utopia/dystopia that sonically locked all present into a state in which the refrains of good and evil alike were subordinated to the will to cast the other as outsider.

AUDIOTOPIA

In conjunction with Williams's premise that 'the social history of religious movements in the United States demonstrates that the boundaries of the culturally legitimate may change, but at any one time they push minority groups firmly to the margins' (Williams 1995, 313), it can be posited that the Davidians came to signify, for mainstream culture, sociopolitical excess—that which is removed from the somatic and psychological context of the everyday, or, in other words, 'waste'. The first question suggested by this notion of excess is what could constitute 'waste' on the frequency-based terrain. Schafer (1970) offers an answer in his insistence that, ever since industrialisation, man-made sound has become so profligate that it can be seen as analogous to sewage and pollution. Schafer is particularly thinking of the problems associated with noise—the excessive frequencies made by humanity that have, in his words, become an 'epidemic': the rumble of cars and their stereo systems, the industrial machinery that repairs roads and produces goods, and, we might add today, digital technologies that blip, bleep and grunt and mobile phones that shriek and roughly caress the ears of those who answer them. For Schafer, all of these sounds are sonic detritus, waste that needs to be placed outside of culture, beyond the periphery of the lived, consigned to the cemetery of those excess frequencies which, he suggests, are always threatening to overwhelm our perception of the here and the now.

Before Schafer developed his theories about noise, English science-fiction author J. G. Ballard had written 'The Sound-Sweep', a short story based in an ultrasonic world where 'noise, noise, noise [is] the greatest single disease-vector of civilization' (Ballard 1960, 52), a scourge that must daily be 'swept up' in a fashion similar to the way we pick up and dispose of garbage today. Evoking the violent cacophony of this future world, Ballard describes the conditions in which the refuse collector of noise must conduct his hygienic operations: 'Occasionally, when supersaturation was reached after one of the summer holiday periods, the sonic pressure fields would split and discharge, venting back into the stockades a nightmarish cataract of noise, raining on to the sound-sweeps not only the howling of cats and dogs, but the multilunged tumult of cars, express trains, fairgrounds and aircraft, the cacophonic musique concrète of civilisation' (ibid.).

The offensive aural detritus Ballard catalogues here significantly anticipates the FBI's set list at Waco while also adroitly commenting on the hostile reverberation of a culture's waveformed output. Ballard both implicitly predicts the utilisation of noise and music as a weapon and preempts the fluid frequency dynamics that allow infrasonic, ultrasonic and sonic content—emanating from the 'echobin' of the city—to be effortlessly channelled into militarised theatres of operations.

From a different perspective again, Jacques Attali considers noise and the notion of waste to have a very different relationship to human perception and existence, stating that 'our science has always desired to monitor, measure, abstract, and castrate meaning, forgetting that life is full of noise and death alone is silent: work noise, noise of man, and noise of beast. Noise bought, sold, or prohibited. Nothing essential happens in the absence of noise' (Attali 1985, 3).

According to Attali, we are intrinsically immersed in and composed by the noises we create, the sonic faeces we produce, the dead we become. We are covered with and embraced within our own decay, our excess, our wilderness—something that was certainly true at Mount Carmel, where 'published reports and survivor testimony indicated that the . . . facility lacked indoor plumbing and that Branch Davidians were required to haul buckets of human waste out of the residential area and to dump them elsewhere on the premises' (Ellison and Bartkowski 1995, 113). This state of excessive decay within the sect's living conditions was echoed and amplified by the acoustic conditions produced by the FBI's speakers, the compound becoming a contradictory site that could be described as an audiotopia—a waveformed version of Foucault's heterotopia in which a sonic domain comes to represent other frequency-based sites within a culture but is itself purposefully temporally and/or spatially displaced.

In the West we understand and measure civilisation by its capacity to engineer and maintain systems that consistently remove what we consider to be excess, waste or dead, away from our everyday living situations. The sites to which they are removed echo Foucault's theoretical constructions of the cemetery, which he refers to as a *heterotopia*—a site mirroring other spaces but removed from the heart of the city—so that the 'cult of the dead' (Foucault 1966, 233) and the reality of decay can be excluded from our cultural gaze. The refuse dump and sewage system likewise serve to deliver the obsolete, the redundant and the excessive away from the spaces of civilisation. All that is waste—that which causes stasis and is considered a burden on the flow of capitalist bodies, information and life—is driven out by elaborate systems of expulsion. On the edge of civilisation, bordering the wastelands and in excess of mainstream values, the Mount Carmel ranch was replete with all the signifying traits of a heterotopia. But insofar as the standoff developed primarily in an acoustic topography, the term *audiotopia* provides a more appropriate way to analyse and describe the complex set of spatial, political and sociosonic relations that evolved during the siege.

As with Foucault's heterotopia, the audiotopia involves an echoing (as opposed to Foucault's ocular 'mirroring') of other sites in the city. When the FBI forced the Davidians to listen to noises from the dentist's surgery, religious music from the church and pop music from the nightclub, an apparently random set of domestic, scientific and social spatialities were brought to bear in the wilderness of the compound. Mount Carmel became analogous to the sonic dump frequented by Ballard's sound-sweep, a site into which all noise and audible refuse is tipped once it has been swept up after a long day's reverberation in the city and the suburbs. Accordingly, the 'occult performance of the state of siege' (Virilio 1977, 36) constructed an audiotopic spatiality—a conflictual and ambiguous sonic space on the edge of civilization where symphonies of conflict were (out)cast by duelling protagonists who understood each other to represent the living (but soon to be) dead.

LOUDHAILERS OF THE LIVING DEAD

Sonically invoking the living dead in order to terrorise an adversary was not a new strategy for the US government. During the Vietnam War the psyops division of the US military had devised a ghostly sonic strategy named 'Wandering Soul', which employed a literal interpretation of haunting to induce an overwhelming sense of angst, fear and anxiety within enemy territories. The Wandering Soul recording—also referred to as 'Ghost Tape Number 10'—was part of the 'Urban Funk Campaign', an umbrella term for the operations of sonic psychological warfare conducted by the US during the Vietnam conflict. After researching the religious beliefs and superstitions of the Vietnamese people, psyops initiated this 'audio harassment' programme, which amplified ghostly voices to create anxiety and fear within resistance fighters. Explaining the rationale behind the strategy, on the *Psywarrior* website SGM Herbert A. Friedman (Ret.) tells us that 'these cries and wails were intended to represent souls of the enemy dead who had failed to find the peace of a proper burial. The wailing soul cannot be put to rest until this proper burial takes place. The purpose of these sounds was to panic and disrupt the enemy and cause him to flee his position. Helicopters were used to broadcast Vietnamese voices pretending to be from beyond the grave. They called on their "descendants" in the Vietcong to defect, to cease fighting' (Friedman 2009).

Blasting frequencies ranging from five hundred to five thousand hertz at an amplitude of 120 decibels from a helicopter-mounted speaker system named the 'People Repeller' or 'Curdler', the US military amplified mournful ghostly voices during the dark hours of the war-torn nights. This was not the technique of the quick kill; rather, it was a technique requiring longevity, slowly diminishing resistance by infiltrating every psychological pore of the enemy's willpower. As Virilio states, the purpose of employing such techniques is that they 'provoke a prolonged desperation in the enemy, . . . inflict permanent moral and material sufferings that diminish him and *melt him away*: this is the role of indirect strategy, which can make a population give up in despair without recourse to bloodshed. As the old saying goes, "Fear is the cruellest of assassins: it never kills, but keeps you from living"' (Virilio 1977, 63; emphasis original).

The location of the speakers in Vietnam—on the sides of helicopters—differs greatly from the surround-sound placement of speakers applied to the conflict at Waco. The more random, transient nature of the amplified wailing sounds had been designed to evoke the idea of 'the restless' being trapped in an environment unsuitable to its noncorporeal status. Rather than scoring the sonic boundaries between states of wilderness and domination as at Waco, in Vietnam the stakes had been raised by framing the border as a tenuous state between life and death.

It is true to say that 'when sound power is sufficient to create a large acoustic profile, we may speak of it, too, as imperialistic. For instance, a man with a loudspeaker is more imperialistic than one without because he can dominate more acoustic space' (Schafer 1993, 77), and yet this does not fully take into account the specific impact of the content. In Vietnam, the sonic demarcation enacted from on high was more of an audio erasure of the boundary between the living and the dead, rendering the absent as distressingly present. The proposed psychology of this strategy suggested slippage and existential echo.

The disquietude of the Vietcong fighter being out of body, place and time suggested that unless they surrender, the only realm conceivable for them was that in which they would be trapped, wandering for the rest of time. For the Vietcong, the airborne sonic virus that was 'Wandering Soul' propagated anxiety and restlessness as it made communicable the oscillating channel of purgatory. Quite literally, it was the sound of hell on earth.

Linking up Vietnam and Texas through the patch bay of history, the concept of the waveformed virus may allow further rumination upon the malevolent amplitude of the spoken word within conflict. In Vietnam, psyops employed the spoken word as a residual frequency of the living dead, whereas in Waco the voice played a more cryptic role, as it was channelled through loudhailers, speakers, phones and radio stations to infect its target audiences. As Koresh desperately tried to transmit to the FBI, the media and the wider public his interpretation of the Seven Seals, the FBI was synchronously attempting to implant doubt and scepticism into his followers. 'There is perhaps also something of William Burroughs' conception of the parasitic "viral" voice in the authorities' recordings and verbal hailings via loudspeaker and telephone', Madsen writes.

> I certainly reference this in my usage of the word 'possessed'. For Burroughs a concrete 'demon word'—literally and materially proliferating, invading and controlling bodies of all kinds via media—could constitute a very direct, although concealed, form of warfare. In his oral story-telling, Burroughs spoke of such sonic weaponry, imagining a new weapon of 'sympathetic vibration' in *The Job*—which 'magnified sound frequencies' to shift 'the battlefield to the internal arena of the body itself'. Is it possible that something like Burroughs' viral demon-like voices and parasitic words were in operation in this sonic assault? (Madsen 2009, 95)

Continuing this line of thought on the power of the viral voice to instigate a throwing down of arms, Madsen picks up on the unsound proclivity of ventriloquism and the throwing of voices to further her notion of the heard and the unheard. For her, these camouflaged acoustic anchors cast in the drowning sounds of Waco were 'noisy ghostly emissions . . . voices able to attach themselves parasitically to other bodies and organs, summoning, possessing and calling out through the voice of another' (ibid., 92). While suggestive of the charnel house that the compound was to become, this explanation of Mount Carmel's sonic terrain both alludes to Vietnam's acoustic purgatory and reverberates with Ballard's description of the sound dump: 'a place of strange echoes and festering silences, overhung by a gloomy miasma of a million compacted sounds, it remained remote and haunted, the graveyard of countless private babels' (Ballard 1960, 61).

NOTES

1. Utilised in combat and peacetime by militaries around the world, *psyops*, or psychological operations, are an important force protector/combat multiplier and a nonlethal weapons system. Often using radio, printed materials, visual media and other loudspeaker techniques, 'psyop[s] are planned operations to convey selected information and indicators to audiences to influence their emotions, motives, objective reasoning and ultimately the behaviour of organizations, groups, and individuals' (Rouse).

2. A good place to start is with the six-episode miniseries *Waco* created by Drew and John Erick Dowdle, which premiered on Paramount Network in 2018.

3. Also referred to as a *model country* or a *new country project*, a micronation is a collective of people (although it can allude to an individual as well) who profess independence and sociopolitical autonomy. In many cases, as noted by Erwin Strauss (1985), no other major international organisations or governments recognise their status as such. Examples include the former World War II North Sea fort now known as the Principality of Sealand.

4. Those who do not identify themselves within—or/and are not accepted within—the social dictates of the mainstream culture or social body they are on the periphery of or outside of. The proliferation of 'outsider' groups that have sought the social margins since World War II has increased, as the static order of the prewar era gave way to a newly contingent organisation of politics, countries, states and territories post-1945. Within this period of fluid dynamics, composed as it was of social reformation, geographical remapping and economic reconfiguration, the relative chaos that ensued within the reorganisation of the world meant that new wildernesses opened up that had not been perceived as such before. As fictionalised with (black) humour by Thomas Pynchon (1975), this postwar state of confusion in which moral, civic and fiscal (and other) boundaries were rendered ambiguous and set in a near constant state of transformation resulted in religious sects, political organisations, cults, ecological groups and micronations seeking their own version of a wilderness where they could protect their sense of agency through self-governance. With the restructuring of international economies, laws and communities that has occurred during the past two decades of globalisation, the circumstances have become once again predisposed to the proliferation of independent (outsider) groups seeking the hinterlands for reasons of sociopolitical autonomy.

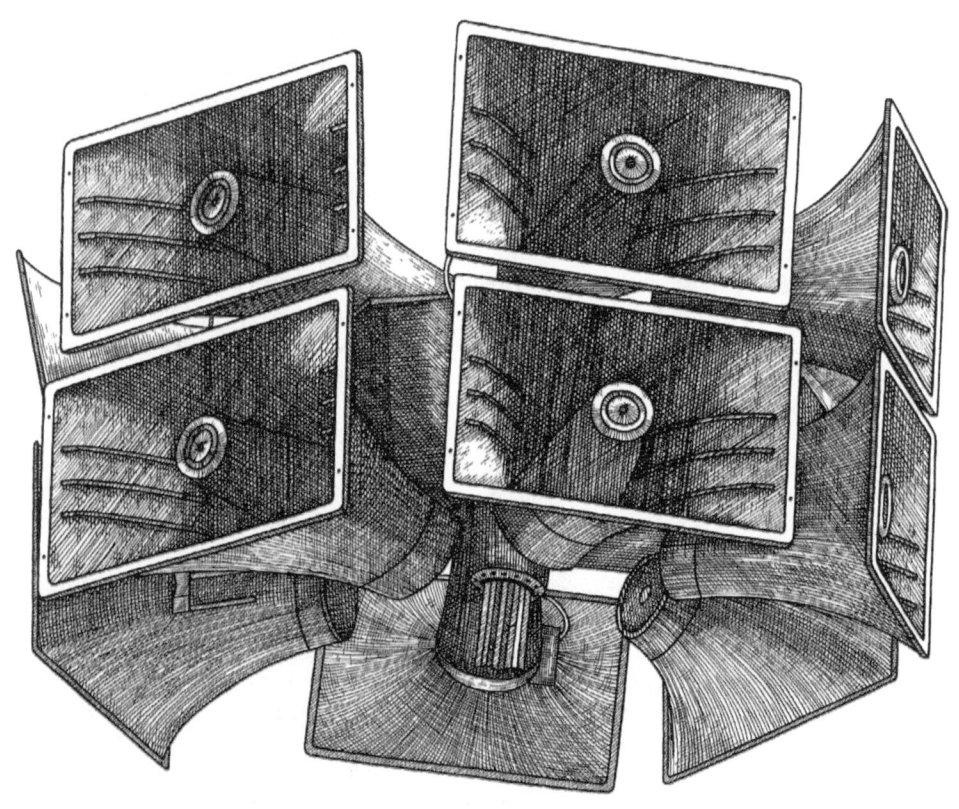

IMLCORP SoundCommander 3500PA Deployable PA System.
Illustration by Krystian Griffiths

Chapter Three

Torture in Black Ecstasy at Guantánamo Bay

SONIC INTENSITY

This chapter concerns the state-sanctioned sonic-torture techniques utilised on persons illegally captured and sequestered within the Guantánamo Bay detainment camp in Cuba[1], which since 2002 has been operated by the Joint Task Force of the US government. More particularly, it considers the waveformed dynamics of violence at work in the isolation/torture cells and the physiological, spatial and psychological status of those tortured within them. If chapters 1 and 2 suggested a theoretical trajectory of instances in which the military-industrial-entertainment complex had isolated numerically ever-smaller groups into progressively more confined architectural circumstances—from the factory and its workforce to the compound and its extended family—this chapter documents the ideological endpoint of this trajectory. With Guantánamo's categorical disproof of Foucault's maxim that 'we are now far away from the country of tortures' (Foucault 1975, 307), here we will contemplate the situation of the isolated detainee sonically immobilised and tortured within a cellular architecture.

As we saw previously, the interwoven sonic and architectural techniques employed to influence industrial workers in the 1920s were psychologically, spatially and physiologically intensified in Waco to manipulate the activities of a trapped and 'compounded' sect. In Guantánamo, these levels of intensification will be amplified again, as we investigate the ways in which sound functions as an instrument for the state persecution of targeted individuals. Guantánamo Bay thus comes to symbolise the conclusion of this vector in the form of an acoustically constructed terminal: the architectural form of the cell cannot be reduced any further in size, levels of acoustic repetition cannot be increased any more and the single detainee tortured in his isolation cell cannot be any further alienated in terms of legal, psychological, physical, spatial or sociocultural dictates. In this sense, the torture cell represents the ultimate dénouement of this waveformed trajectory of techniques—a set of practices that can be more clearly defined as the state's capacity to render pure *sonic intensity*.

The practice of torture techniques in Guantánamo culminated in abnormally high levels of self-harm and suicide being recorded as many detainees attempted to escape their 6.5- by 8-foot cells via the only route left open to them—death by their own hand. Examples of such self-inflicted violence are recounted in reports made by Amnesty International, which disclosed in its *Guantánamo Fact Sheet* that in 2003, 350 separate incidents of self-harm occurred in the camp, while in 2005, 110 incidents of self-harm or suicide were recorded and released (Amnesty International 2007). In 2006, after three camp detainees were found dead, camp commander Rear Admiral Harry Harris issued a bewildering statement, declaring that these deaths ought not be seen as an act of despair over conditions in the camps but instead be understood as 'an act of asymmetrical warfare waged against us' (BBC News 2006). As a consequence of mounting national and international pressure, on 15 December 2009 Barack Obama issued a presidential memorandum: 'Closure of Detention Facilities at the Guantánamo Bay Naval Base', in which he ordered that the Thomson Correctional Center in Illinois be readied for the transfer of detainees (White House 2009). Strong resistance from both Republican and Democratic senators to inmates being transferred to prisons in their states meant that Guantánamo was not closed, with Obama's successor, Donald Trump, more recently declaring that the camp would remain open indefinitely. Since its opening, the facility at Guantánamo has received 779 detainees, of which only a very small number have been formally charged with offences. According to a *Times* [of London] report by Tim Reid (2010), while the majority of detainees have been 'moved', we do know that as of August 2018 there are still forty left at the facility awaiting transfer.

Although numerous torture techniques and acts of inhumane cruelty must be taken into account when analysing the full spectrum of Guantánamo's organised violence, this chapter focuses on 'the use of [a] kind of audio-technique [that] is rather new in interrogation', according to vice president of the US Army Psychological Operations Veterans Association, Rick Hoffman (BBC News 2003a). We are speaking of the use of music as torture, which appears to have been widespread throughout all the camps within Guantánamo, reports from ex-detainees being numerous and detailed. According to one report, 'Shafiq Rasul, of the "Tipton Three"—British Muslims detained in Guantánamo for over two years after being captured by the Northern Alliance in Afghanistan—tells of being short-shackled to the floor in a dark cell while Eminem's "Kim" . . . and pounding heavy metal played incessantly for hours, augmented by strobe lights' (Hultkrans 2008).

It was via the reenacted scenes of Michael Winterbottom and Mat Whitecross's 2006 documentary drama *Road to Guantánamo* that this recently developed sonic technique firmly entered the public imagination, broadcasting the searing acoustic brutality of sensory overload into our collective consciousness.

In an interview for *Der Spiegel* with Shafiq Rasul's detained friend Ruhal Ahmed, German writer Tobias Rapp leaves us in no doubt as to the serious psychological threat of this sonic practice that has become part of the arsenal of so-called torture lite[2]. Ahmed tells Rapp, 'You can't concentrate on anything. Before that, when I was beaten, I could use my imagination to forget the pain. But the music makes you completely disoriented. It takes over your brain. You lose control and start to hallucinate. You're pushed to a threshold, and you realize that insanity is lurking on the other side. And once you cross that line, there's no going back. I saw that threshold several times' (Rapp 2010).

While Ahmed goes on to reveal how he was short-shackled for days at a time and 'left to urinate or defecate in his pants' in ice-cold chambers of sonic torment, it is his account of the inversion of music's capacity for psychologically transporting the listener that is most revealing: '"When you go to a concert or a club, you're looking for loud music and flashing lights. You want to be transported into ecstasy. We experienced exactly the same thing, except that it was turned on its head", says Ahmed. "You could call it black ecstasy"' (ibid.).

Rapp's article, along with campaigns such as Zero dB[3], exposes the state's employment of popular music by artists such as Eminem and Metallica and other selected 'mainstream' materials for use in the application of force. Jon Ronson's interview, meanwhile, with Jamal Udeen Al-Harith (another Briton held in extrajudicial detention as a suspected terrorist who, while never charged, was tortured—hence his involvement in the Rasul v. Rumsfeld case[4]) reveals a different aesthetic approach to sonic torture altogether. In conversation with Al-Harith about the types of music used to disorient and disturb him in Guantánamo, Ronson cannot conceal his surprise at learning that guards played atonal soundscapes that had no beats or rhythms, aural collages that could be termed *experimental electronic compositions*, consisting of noise, industrial sounds, electric piano and synthesiser lines (Ronson 2004). While we may have become used to reading reports of current popular music genres such as country, heavy metal and rap being employed by the military for their abusive day-and-night torture sessions, the revelation that sonic compositions plausibly comparable with avant-garde cultural production were used seems bizarre. But is it in fact rather telling?

While musical genres such as country and rap enjoy a huge market share and attention in the United States, they also have traditional structures at their core that have long historical lineages and that often share narrative interests and sociopolitical affiliations with musical genres from other countries. Country music, for example, shares traits with traditional storytelling modes of music such as the European sea shanties of the nineteenth century and mariachi music from Mexico, while we might trace the rap genre back to its West African traditional oratory roots in the songs and poems of griots; or, thinking of more lateral associations, we may point to the attitudes that resistance rap has in common with Afro-Brazilian musical forms, martial arts and dance known as capoeira, realising that the lineages of these musical genres have enduring and identifiable connections to global forms of musical organisation and politicisation.

Avant-garde music can plausibly be considered to be more deeply entrenched in Occidental value systems than can either of these mainstream American genres. This means that avant-garde music can be understood to represent the culture whence it came in a more nuanced and deeply rooted way. For those forms that speak most intimately about a culture and its particular distinguishing characteristics, those phenomena that set one culture apart from another (and thus help construct identity) are surely not the mainstream ones. Surely they are not those genres or movements of people, ideas or practices that generate the amorphous and ambiguous models of globalised culture that speak about everywhere and tell us everything about nowhere. As Jacques Attali writes, 'Mass music is . . . a powerful factor in consumer integration, interclass levelling, cultural homogenization. It becomes a factor in centralization, cultural normalization, and the disappearance of distinctive cultures' (Attali 1985, 111).

Instead of thinking about the weaponisation of mainstream and global cultures, in this instance it is useful to consider the assimilation of grassroots activities, micromovements and the modulations of smaller DIY and avant-garde cultures for such purposes. Musically speaking, it is just such newly emerging, experimental and radical sonic gestures that better articulate site-specificity, psychological associations with place and sociocultural identity. As a genre, we could say that the avant-garde is therefore more definable as idiosyncratically Western—and a fortiori American—than any mainstream genre we might think of as sonically symbolic of the United States. Following this logic, avant-garde productions can be posited as effective signifiers of specific cultural identity, a proposition that the military-entertainment complex may well have been exploring given its history of cultural assimilations for martial purposes that have been explored previously in the text, and which will be examined further in this chapter. In a more idiosyncratic version of this avant-garde exploration, eschewing the need for prerecorded music of any specific genre with which to inflict damage upon detainees, British soldier Donald Payne orchestrated his own perfidious form of musique concrète (for which he was jailed). This was engineered by conducting what he called 'the choir': striking the Guantánamo prisoners in sequence, their groans or shrieks making up the 'music' (Johnson and Cloonan 2009, 157).

At the other end of the musical spectrum, the genre of disco figures heavily in the analyses both by American musicologist Suzanne Cusick and by English professor Moustafa Bayoumi of US torture practices in Iraq. In his article 'Disco Inferno' for a December issue of *The Nation*, Bayoumi investigates the torture rooms in Mosul, writing of detainee Haitham al-Mallah who 'described being hooded, handcuffed and delivered to a location where soldiers boomed "extremely loud (and dirty) music" at him. Mallah said the site was "an unknown place which they call 'the disco'"' (Bayoumi 2005).

In his conclusion, Bayoumi surmises that 'torture threatens to decivilize us today not only because its practices are being normalized within our national imagination but also because civil society is being enlisted to rationalize its demands' (ibid.).

This notion of disco being 'dirty music' is not a new narrative in the West. As noted by Richard Dyer (1979), disco has commonly been associated with its roots—1970s clubs whose clientele consisted mainly of African American and homosexual communities who supposedly practiced sexually deviant behaviours and embraced excessively hedonistic lifestyles (often involving drug taking). Cusick further examines the utilisation of music that has been labelled 'queer' or 'effeminate' by champions of overtly masculine, heterosexual culture. She proposes that such types of music are employed as a weapon because of their perceived association with and promotion of excess and deviancy, which in turn are understood to be transmissible qualities that can infect the religiously pure body of the detained male Muslim (Cusick 2006).

Examining bloggers' responses to the notion of music as torture, Cusick touches upon this cultural paranoia about 'dirty' types of music that are understood to soil 'clean' Western value systems. In her 2006 article 'Music as Torture/Music as Weapon', she comments that a number of bloggers 'use the idea of music as torture to displace onto Muslim detainees a rage rooted in their own fear that they are immersed in a culture that has become, in their words, "nancy", "pansy" and "pussy"' (ibid.).

For Cusick, musical genres such as disco are employed against the incarcerated bodies of Guantánamo to disturb, contort and distress them via sexual connotation and transference; the cell's acoustic organisation (the disco music) and its efficacy internally ratified precisely because, to the US military officer, it represents a sonic currency of carnal hostility deployed against the detainee.

DISCO INFERNO

Endowed with a momentous capacity to culturally offend en masse, in the 1970s disco inadvertently gave rise to genre-based musical warfare within popular culture. A good example of this conflict is the infamous Disco Demolition Night at Comiskey Park in Chicago in 1979, where between baseball games a crate of disco records was blown up by fans of rock music. During the same decade, sonic torture began to be utilised against prisoners in numerous countries. Drilling down into this stratum of acoustic torment, Darius Rejali discloses that in Rio in 1976 prisoners such as José Miguel Camolez were locked in small soundproofed cells measuring two by two metres, called *geladeiras*. These cells contained speakers attached to the ceiling through which the guards used to 'call him dirty names'. As soon as the voices fell silent, they 'were replaced by electronic noises so loud and so intense he could no longer hear his own voice'. Camolez goes on to tell Rejali how the sound then stopped, and the walls of the room were battered '"with great intensity for a long time with something like a hammer". Others mention that the *geladeira* was "very cold", had a "very strong light", and produced varying sounds from "the noise of an airplane turbine to a strident factory siren". . . . An official DOPS record from the period states that the *geladeira*'s objective was the "destructuring" of the captive's personality' (Rejali 2007, 365).

The familiar tone of this historical report makes it seem as though the *geladeira* could have provided the original blueprint for the sensory deprivation/overload techniques of Guantánamo Bay, but this is not the case. For this egregious soundmark to be traced back to its inception requires us to attend to accounts of the British Armed Forces' actions in Northern Ireland during the early 1970s; for it is there that we hear of the first pansensory approach to forcing the body, the legal system (via the third and fourth Geneva Conventions[5]) and the mind into submissive and irrational relations with its captors, via the 'five techniques'[6].

In 1971 Jim Auld, along with a targeted group of Northern Irish Catholic men, were picked up by the British government—without evidence to associate them with any crime—and, as John Conroy relates, subjected to 'a scientific combination of tortures'. A hood was placed over Auld's head, after which noises were played at an intense level. 'Various survivors described it as the sound of an airplane engine, the sound of compressed air escaping, and the sound of helicopter blades whirring. For a solid week, the noise was absolute and unceasing, an assault of such ferocity that many of the men now recall it as the worst part of the ordeal'. Placed into stress positions, the detainees were also denied water, food and sleep and were refused visits to the toilets, meaning that they had to urinate and defecate in the boiler suits they had been forced to wear. A new era of torture had begun: The 'five techniques'—as they were dubbed at a later date—'induced a state of psychosis, a temporary madness with long-lasting aftereffects' (Conroy 2000, 5–6).

These early pansensory experiments informed sonic-torture techniques employed globally—in soundproof, acoustic-abuse rooms in Mexico, in white-noise rooms in Mali Alas in the former Socialist Federal Republic of Yugoslavia and in isolation cells in South Africa soundtracked by tapes playing back the sounds of 'human screams, breaking glass, barking dogs, and roaring lions' (Storr 1992, 101). Sensory-deprivation and sensory-overload techniques had been employed before 1971, but not in the refined version of the five techniques developed in Northern Ireland and later applied in Guantánamo Bay. Rejali informs us of reports made in 1957 by Harold Wolff and Lawrence Hinkle establishing that 'isolation, sleep deprivation, nonspecific threats, depersonalization, and inadequate diets placed enormous stress on individuals' (Rejali 2007, 373). Cusick, meanwhile, reports that 'in 1963 in the CIA's *KUBARK Counterintelligence Interrogation* handbook, the techniques of "no-touch torture" were used—indeed, consciously tested again and again—by the CIA's counterinsurgency forces in Vietnam into the 1970s' (Cusick 2006).

Echoing sonic techniques employed by military units in Guantánamo, Croatia and Northern Ireland, programmes of waveformed punishment have been filtering back into the global civilian fold over the past decade, amplifying once again the overt mixing (development), compression (intensity) and mastering (control) of acoustic technologies and strategies between martial and civil contexts. In 2004, upon sentencing the driver of a car who had been blasting rap music from his car at 5 A.M., a judge in Florida gave an ultimatum: either pay a fine of $500 or listen to two and half hours of Verdi's *La Traviata*. 'You impose your music on me', the judge scolded, 'I'm going to impose my music on you' (Johnson and Cloonan 2009, 176). This is not a lone case. In his 21 January 2009 *Los Angeles Times* article 'He Writes the Rules That Make Their Eardrums Ring', Matthew Staver reports that Colorado Municipal Judge Paul Sacco has been conducting his own symphony of sonic justice called 'music immersion', which works along similar lines (Staver 2009). By effectively collapsing the acoustic walls between the military torture cell and the civilian courtroom, our culture expresses its desire and its capacity to crossmodulate between realms the rationalisation, acceptance and propagation of punishment techniques through institutionalised fear. If this is the case, then it is only through deciphering levels of cultural compression that the nature of the 'enemy' can be defined.

Cultural compression refers to the modification of intensity levels that occurs when technologies are transferred from the battlefield into civilian environments. A prime example is the Long Range Acoustic Device (LRAD) technology, which has been used in both civilian and military contexts. The manner in which it is used will more likely tell us how much of a threat the enemy is perceived to be. Thus the frequency and volume levels applied in situations in Fallujah, Iraq—where 'music was played so relentlessly . . . that the Marines nicknamed the city "LalaFallujah"' (Pieslak 2009, 84)—were not applied in New York when police used LRAD against protestors there. While the techniques and technologies deployed across the sociogeographic realm are the same, the levels of intensity levelled against the targets are modified. Similarly, the rates of repetition and the levels of sonic force of music used against a detainee in a Guantánamo Bay cell are not the same as those employed by judges in the United States to punish those who have violated city noise codes. The point here is that there are levels of compression occurring between martial and civilian usage, but it should be understood that they are malleable and can be easily modulated at any time.

Within Western civilian networks of waveformed relations, systems of sonic punishment appear to clearly operate with high levels of legalised distortion. As the faders of ambient social violence are ineluctably pushed up, they incrementally amplify our willingness to accept that anyone and everyone can become an admissible target. Studying the wave of reports produced by academics and journalists on instances in which sound has been used as a weapon in some capacity reveals a curve of social penetration that arcs between (and consequently connects) the torture rooms of Guantánamo Bay and the teaching rooms in English schools. If this seems far-fetched, let's consider a few examples of the ways music has been utilised as a form of cultural artillery on groups of people who represent a sliding scale of threat—from those manifestly labelled as guilty external enemies to those considered innocent domestic citizens.

This acoustic survey begins by marking the extreme end of the curve, with a note from Bayoumi, explaining how the initial diffusion of sonic-torture techniques from detainment camps transfers them effortlessly into the civilian milieu via the organisational vehicles of bathos and humour. Bayoumi states that music as a form of 'torture lite slides right into mainstream American acceptance. It's a frat-house prank taken one baby-step further—as essentially harmless, and American, as an apple pie in the face' (Bayoumi 2005). He suggests here that the way in which the transferring of a punishment-based modus operandi is negotiated—from the theatre of operations to the movie theatre—is essential to its chances of being able to infiltrate all levels of the social spectrum. When compliant accomplices such as Stevie Benton (bass player from nu metal group Drowning Pool) assist in easing the acoustic cargo through the martial-social border with easy-listening statements that blithely inform us he 'can't imagine it's that bad. . . . Listening to loud music for a few hours—kids in the US pay money for that' (cited in Rapp 2010), this must surely please militant exporters. For there is no more effective method of camouflaging the violent and insidious nature of a technique imported into new circumstances than by wrapping it up in humour and light-hearted commentary—a dynamic that suits both the entertainment and military industries well as they colegitimate their presence in each other's spatial, psychological and economic territories.

Once the transposition has been legitimised by mediated figures such as Benton or former host on Fox News Channel Bill O'Reilly, who 'claimed that if blaring music at someone is torture, the phrase has become meaningless and that to take such a view was "just nuts"' (Johnson and Cloonan 2009, 191), such sonic techniques are readied for civilian use. The uses of music as punishment in the judicial structuring of censure against those who violate the laws of the social system become especially relevant here. The next point on the curve indexes the use of music against those deemed undesirable and outside of the mainstream, such as the homeless and drug addicts: 'Authorities at the main railway station in Hamburg have used piped-in classical music to drive away junkies from the plaza in front of the station' (Rapp 2010); 'Authorities in Copenhagen used Bach to drive drug addicts away from the city's main railway station' (Johnson and Cloonan 2009, 184); 'Stoke-On-Trent Council played Beethoven's Ninth Symphony continuously in car parks to deter rough sleepers. St. James Church in Carlisle used music by Bach and Handel to deter drinkers who had been gathering on its steps and vandalizing church property' (ibid., 183–84).

It is already obvious where this curve is leading us, but there are still a number of points to plot before it can be declared that sonic techniques of manipulation and torture (lite) have become culturally normalised in the West. As documented by Dick Hebdige (1979), following World War II, working-class youths from England and the United States began to form subcultures based on musical affiliations, purposefully marginalising themselves from mainstream culture in the process. The formation of the first 'Teddy Boy' gangs in London in the 1950s cast the youth of the United Kingdom under suspicion, where they were considered a threat to 'civilised' society. In line with such cultural paranoia about the loss of control over younger generations—particularly those that can be associated with musical affiliations decreed to be 'violent' and 'antisocial'—the next point on the curve is exemplified by a report emanating from Holywood—in Northern Ireland—where 'local businesspeople encouraged the council to pipe classical music as a way of getting rid of youngsters who were spitting in the street and doing graffiti'. Cataloguing this and other examples in his 2010 article 'Weaponizing Mozart', Brendan O'Neill reveals the entrenched neurotic attitudes held by many towards those whose greatest offence, it seems, is being young and therefore deserving of sonic harassment: 'Tyne and Wear in the north of England was one of the first parts of the UK to weaponise classical music. In the early 2000s, the local railway company decided to do something about the "problem" of "youths hanging around" its train stations. The young people were "not getting up to criminal activities", admitted Tyne and Wear Metro, but they were "swearing, smoking at stations and harassing passengers". So the railway company unleashed "blasts of Mozart and Vivaldi"' (O'Neill 2010).

We arrive now at the teleological destination of this arc of acoustic violence tracing out the exportation of sonic-warfare techniques from externalised military contexts to everyday urban reality. It is O'Neill, once again, who supplies the final example. He writes: 'In January [of 2010] it was revealed that West Park School, in Derby in the midlands of England, was "subjecting" (its words) badly behaved children to Mozart and others. In "special detentions", the children are forced to endure two hours of classical music both as a relaxant (the headmaster claims it calms them down) and as a deterrent against future bad behavior' (O'Neill 2010).

O'Neill points out that, while these kinds of reports may not shock us, the details resonate all too closely with the dystopia of Anthony Burgess' 1962 book, *A Clockwork Orange*: sensory overload and sonic entrainment as legitimised forms of scientific-cultural discipline, ensuring that we feel as relaxed as the children apparently do after a 'special detention'.

But then maybe the production of fear or urban dread concerning impending techniques of state power is the raison d'être of a military-entertainment complex that aims to encompass us all in the paranoid, intimate embrace of pure war. For Paul Virilio (1977), this is an omnipotent ecology that fosters a constant state of neurosis and preparedness. It works to immutably mesh together the civilian and martial fabric of cultural, economic, social, spatial and technological networks so that they become evermore entangled and even symbiotic, their interests indefinable from one another. Ultimately this creates a situation whereby 'it is impossible to tell where the civilian sector begins and where the military ends' (DeLanda 1991, 228). Within this collapsed logic of difference, new nebulous and malleable spatialities such as Guantánamo Bay are opened. Sequestered in such a legally and ethically dysfunctional orbit, it means that 'the Guantánamo detainees are located in the space "between the two deaths", occupying the position of *Homo sacer*, legally dead (deprived of an official legal status) while biologically still alive' (Žižek 2006, 371).

ASYMMETRIC CONFLICT

As the term suggests, the *military-entertainment complex* is not easily definable, and this is where its power of *détournement* resides; having become integrated and harmonised, the spatialities, psychologies and rhythms of leisure and conflict find their consummate expression of oscillating versatility in the frequency-based domain. The previously verticalised divisions—the walls of sound—that used to separate the cultural from the martial, the city from the battlefield, the song from the torture weapon, have been erased over time, rendering a tabula rasa of violence that has flattened the world out into resonant latitudes of conflict. In this scenario it is not difficult to connect the paranoia that fuels the persecution of hooded youths in the United Kingdom with the black-hooded Muslims in Guantánamo Bay, even if the levels of sonic intensity used on the two groups are very different. Taking this a step further, the two groups can also be associated through their increasing representation in films, video games and music. Indeed, this opens another dimension of the exchange—namely, the entertainment industry's persistent hunger for new violent subject matter that can be converted into digital audiovisual formats and sold for public consumption. Under the rubric of 'real life', marketing agencies sell cultural products containing intense content, on the grounds that they are conveying an authentic experience of the painful difficulties of existence (for example, musical genres such as gangsta rap, video games such as *Grand Theft Auto* [Rockstar Games 1997] and war films such as *Jarhead* [Mendes 2005]).

A mutually beneficial relationship plays out in the realm of the military-entertainment complex in which the two industries exchange and codevelop media technologies and, perhaps more importantly, also produce content to play through them. If this is the case, then does war need music just as much as music needs war? A rhetorical question, but one that sets the correct tone when enquiring whether, through its capacity to become entertainment, violence has become a dominant, endlessly exchangeable form of Western cultural currency. In Guantánamo Bay the music used to torture detainees, to prepare areas for conflict and to boost soldiers' levels of aggression is invariably drawn from the metal or rap genres, both of which have strong affiliations with masculine narratives of alienation, violence and retribution. That mediated representational violence in the form of music is necessary to engineer more real violence on the ground appears to support the notion that this cognitive mechanism is an implicit element in the twenty-first-century engineering of conflict and brutality. As Steve Goodman notes, 'When the most banal popular music is simultaneously mobilised as a weapon of torture, it is clear that sonic culture has reached a strange conjuncture within its deepening immersion into the environments of the military-entertainment complex' (Goodman 2009, 190).

As the ambient tides of urban violence rise, so mediatised representations of current conflicts increase the sonic and graphic calibre of their echoes and reflections. As we become increasingly immune to represented violence, it takes more-elaborate and -extreme forms of mediated brutality, disorder and cruelty to move us to click ADD TO CART. The military-entertainment industries know this only too well and excel in developing and upgrading (1) martial technologies that can cause novel and escalating forms of damage and pain, (2) leisure technologies capable of capturing, augmenting and digitally reconstructing these intensifying techniques of violence and (3) economically viable distribution networks that form feedback loops between the technologies that cause, mediate and represent violence and those who operate them. One could go so far as to say that, just as it has been suggested that the Iraq conflict was a war over the control of oil production, it could also be argued that Iraq and subsequent conflicts were born of the craving for new sonically and graphically violent content for the entertainment industries.

This may seem overstated until one considers the worldwide box-office grosses for those films directly based on the Iraq War alone. A brief glance at Wikipedia informs us that in 2018 there were sixty-three Iraq War documentaries and forty-eight Iraq War films, with the 2010 Oscar-winning phenomenon *The Hurt Locker* being the most obviously economically successful (*Wikipedia* 2018a and 2018b). In relation to the music industry, a burgeoning music-production culture has developed from soldiers writing, recording and mastering their own albums while in Iraq or after a tour of duty. The lyrics are predictably graphic, purporting to bring the reality of life on the streets of Baghdad to those prepared to listen. With monikers and album titles such as Gorriors of Ragnarok's *Rape Guts* (2007), Peaches and Kim's *Escape* (2007) and the 4th25's *Live from Iraq* (2005), there is also no shortage of grassroots productions beneath the surfeit of commercially released mainstream records commenting on the Iraq War in some manner. For computer-gaming culture, there is *Kuma\War* (Reality Games 2004) with its promise of 'playable re-creations of real war events'. All of these industries boast a massive array of productions alluding to the war in Iraq; these few examples merely serve as locating devices to help map the ubiquitous presence and profitability of violence in the conflicted topography of our martially entertained culture.

In 1812, conflict simulation was first used when rules were created for *Kriegsspiel*—a war game used to train officers in the Prussian Army—which is attributed with having had a defining influence on the surprising victory in the 1871 Franco-Prussian War of the German states over the Second French Empire. Ever since then, war games have played an important role in our political and entertainment cultures, and the reality-based and interactive nature of such games seems more relevant now than ever. For the gaming, music and film industries, only war can provide us with the constant feed(back) of horror, violence and shock that will satiate a public with a voracious appetite for witnessing 'real' violence. The fact that soundtracks from computer games based on the conflict in Iraq end up being utilised in torture cells in more intense and repetitive ways is no longer shocking, nor is its obsidian irony even seen as particularly worthy of comment. This is not a case of simulation so much as one of reciprocation—that is, a repeating motion. It is the aim of the military-entertainment complex to create disequilibrium, to displace the simple harmonic oscillation of the body and unbalance it, that completes the feedback loop, from the shooting, screaming (C)GI to the rhythmical mutiny of detainees in the Guantánamo torture cells; and it is the movement of, and into, the body that shall be considered next.

BREAK IT UP, BREAK IT UP, BREAK IT UP, BREAK DOWN

In the attempt to break the detainee's body and mind, music is the most effective technique of no-touch torture[7]. Analysing the relations of power implied in the use of sonic weapons, Cusick asks, 'What better medium than music to bring into being (as a felicitous performative) the experience of the West's (the infidel's) ubiquitous, irresistible Power?' (2006). Conducting research along similar lines, Ramona Naddaff suggests that 'music torture impels a rethinking of how musical listeners experience being touched by sounds that harm their inner senses and maim identity formation' (2009). Such a rethinking would require us to investigate the relationship between haptic spatiality, the socially constructed body and the waveforms that fuse, modulate and separate them. In Guantánamo, following this thirded analytical principle to its logical conclusion leads to an archetypal presence that is innately connected to itself and to the other through a politics of oscillation; in its capacity to synthesise, to have impact and to be impacted upon, the 'body is rendered as multi fx-unit, as transducer of vibration as opposed to a detached listening subject isolated from its sonic objects' (Goodman 2009, 46).

The sheer weight of the sonic mass pressuring the body in Guantánamo was generated precisely to control and possess the culturally compressed anatomy of the other. The alterity and solipsistic spatiality of the torture cell reverberated with echoes of the self, the 'enemy' and the architecture, the divisions between them collapsed by the waveforms that constitute sonic dominance. For Julian Henriques, 'sonic dominance is visceral, stuff and guts. Sound at this level cannot but touch you and connect you to your body' (Henriques 2003, 452). Skin, hair, nails, sinews and muscle all pitch, ripple and swell as the sound waves pound upon, into and through them, the 160-decibel[8] vibrations being received via 'bone conduction as well as through the acoustical properties of the air' (Ihde 2003, 66); it is within this coercive ocean of sound that we find the unwilling antenna-body channelling all waveforms, even those in which it is supposed to drown.

For, even in such submissive circumstances where a dynamics of overloading the organs of perception predominates, the antenna-body still transmits, albeit through noisy channels of interference; it speaks of the pain implicit within repetition and of the struggle for sentience amid overwhelming acoustic intensities; it reveals the 'dark ecstasy' (Cusick 2006) that resides within the 'will' to secure complete dominion and direction of waveforms in space; and it tells us that, no matter how severe techniques for the silencing of difference become, it will always communicate and reveal the structures, techniques and strategies that try to deaden it. With its capacity to report the intensities, dynamics and constructions of military/civic power relations, the antenna-body articulates itself from the cell next to Henriques' 'sonic body': 'This is a whole resonating, specific, shared, social, immediate and fleshy body. . . . The term "sonic body" implies either and both the body of the sound and the sound of the body' (Henriques 2003, 471).

In the cellular cacophony of Guantánamo there are scores of other types of waveformed bodies like these, requiring identification, patiently waiting for the opportunity to inform us of their pan- and cross-sensory characteristics, which may lead to more nuanced conceptions of what it means, physically, spatially and psychologically for the somatic to bear intense sonic pressure. As compelling as this narrative of extreme relational sonic embodiment is, it also needs to be acknowledged that, while the music in Guantánamo connects and transforms the body, the air and the architecture of the cell, an inverse operation of waveformed disembodiment is simultaneously being carried out[9]. Music is particularly effective at negotiating and performing this procedure because, 'like touch, sound has this other opposite aspect that separates us from ourselves, each other and the world. . . . Just as the tactile sense is preeminent in determining the organism's simultaneous connection with and separation from its environment, the sonic sense plays a similar combining and separating role' (Henriques 2003, 461).

By understanding what is entailed in this sonic disembodiment—what its routines and procedures are—we can better comprehend what it ultimately strives to achieve. This will also reveal a great deal about those with a remit to acoustically make manifest that which is deemed 'hidden' and about others who seek to transmute (the detainee's) silence into something knowable, recordable and effable. Commenting on the historically legalised nature of torture, Foucault surmises that its legitimisation as a practice was a result of its capacity to render observable results. He writes, 'It revealed truth and showed the operation of power. It assured the articulation of the written on the oral, the secret on the public, the procedure of investigation on the operation of the confession; it made it possible to reproduce the crime on the visible body of the criminal' (Foucault 1975, 55).

In Guantánamo, reproducing the crimes on the detainee's visible body is problematic, not least because there are no proven crimes whose brutality might be metaphorically reciprocated in this way. Here, instead, the state's desire to embody its revenge by carving its abstracted logistical initials onto the anatomy of the detainee had to be inverted: no traces would be left on that which could be observed. This new disembodied torture practice would require means by which to invisibly score *into* the body rather than onto it, which is why music is so effective in such circumstances. Music does not leave marks because it does not operate by merely touching or representing its power on the somatic interface; it is instead committed to enveloping the anatomical surface, moving into and beyond it, challenging the rationale of the perceivable and quantifiable: 'In fact with sound it simply does not make sense to think of having an inside and an outside in the way that the visual sensory modality, with its preoccupation with surfaces, restricts us. Sound is both surface and depth at once' (Henriques 2003, 459).

EXCAVATION, AUTOPSY, EXORCISM

The potential of waveforms to transgress surfaces has been touched upon already, but we need to up the ante when trying to verbalise the concerns of a body that is simultaneously being sonically punctured and saturated. Hence, the shift in tone in what follows as we try to explicate the sonic turbulence and tumult that is impacted and compressed upon, into and through the hollowed-out figure of the torture cells. Rather than analysing how anatomies are touched, the text will now investigate how the body is breached, penetrated and emptied. In doing so we ultimately hope to reveal how the somatic is brutally operated upon by sound in order to locate discrete and fictionalised forms of knowledge allegedly concealed within it. Here we will consider three disparate methods employed by divergent ontological and epistemological practices to seek out hidden and esoteric phenomena: *excavation*, *autopsy* and *exorcism*.

The first and possibly most abstract way of thinking about how the body is plotted, entered and searched is in terms of excavation. This is a useful concept when thinking about how the body is searched in relation to the organising spatiality of the architectural cell. The body trades places with the architectural walls of the cell, becoming itself the material container understood to potentially contain secrets and useful information. Like a wall, the will is perceived to be something that can be taken apart and broken. If it is correct to say that 'the process of embodiment can only take place through the sensory perceptions' (Henriques 2003, 466), then it is also accurate to say that when the organs that provide those sensory perceptions are overstimulated, they become synonymous with entry points to the body (for the organs, as well as informing the self, can soon be made to betray it). It is from these inlets that sonic shafts are driven and the disembodying processes of mining information begins. This unearthing and tunnelling of the body expresses the state's urge to discover the essential articles of faith that define the 'unknowable' culture it is investigating—to expose its organising principles and to define the (metaphorical) objects of desire, disgust and sacrifice of that which it does not understand, in order to overcome its legacy.

As noted above, for those who confine them, the Guantánamo detainees are already considered legally dead, so an acoustic autopsy of the living is the next logical step in such an irrationally mandated environment. As the strobing lights flash down on the body, they highlight the reactions of the detainee as his anatomy and mind are scored by repetitious orchestrations. The motivation for such a procedure is to find reasons why the detainee has become (legally) dead, for the detainees have been immediately found guilty and sentenced upon capture, without meaningful legal or ethical recourse. The music is released into the cells and each song 'listened to anesthetizes a part of the body' (Attali 1985, 111). While the mind is numbed and emptied, sonic scalpels slowly attempt to remove it, neuron by neuron, note by note. Pain is inevitable when such a procedure is carried out upon the living, emptying the body of its contents in order to identify the causes of deaths—both those he is presumed to have caused and his own drawn-out demise.

Torture and punishment routines are often epitomised by their will to reveal knowledge considered to lie beyond the physical and social body. With this in mind, we will now contemplate the third technique for searching the body, the aural exorcism. For centuries, the kind of knowledge labelled as being beyond both the collective and individuated corpus has changed depending upon the state's desire to reveal or expose some hidden or threatening phenomenon. In the case of the social body, the United States has a history of paranoia and neurosis concerning hidden internal threats. Such anxieties were made manifest through the search for clandestine socialist sympathisers in the 1950s, for example, who were legally punished for their beliefs during the era of McCarthyism. In *Discipline and Punish*, Foucault discusses a different type of existential search into and beyond the physical body of the individual, one that discloses the judicial system's preoccupation with accessing the prisoner's essential and graspable seat of consciousness. He writes, 'If the penalty in its most severe forms no longer addresses itself to the body, on what does it lay hold? The answer of the theoreticians—those who, about 1760, opened up a new period that is not yet at an end—is simple, almost obvious. It seems to be contained in the question itself: since it is no longer the body, it must be the soul' (Foucault 1975, 16).

In Guantánamo, the rationale of sonic torture is not to connect with the prisoner's soul so that the state might try to retrain it and realign its offending compulsions with the rhythms of its own value systems. Rather, it is to expose what the state perceives to be the nexus of evil that resides within the very core of the detainee and that can only be communicated with by repossessing the prisoner's inner voice. In order to break the detainee and locate this inner language of evil, the inner language must be contacted and seduced from its quiescent interiority, to speak and reveal its covert intentions. The interrogators who religiously carry out the torture have to believe that there is a fundamentally monstrous nature within these secured bodies, an unnatural power that endows them with the appearance of the living.

In the torture cells, no religious incantations are reiterated so as to aggravate and draw out the force of evil, for the speakers here are not priests; they are technologies of a different repetitious rhetoric—namely, a form of music that has itself been accused of worshipping and summoning evil: heavy metal. In a darkly ironic turn, the Occident has awoken and amplified the devil through music and is now directing the resulting sonic terror at the detainee, bringing pressure to bear upon the resistor's will to silence. As previously intimated, the spectre of the devil has historically been conceived of as having a particular affinity to sound and unsound. For example, 'the early Christian church believed that pagan residues in music could be exploited by the Devil to produce depravity' (Johnson and Cloonan 2009, 32). Such a proposition suggests that the darkly adumbrated form of the demon could only become embodied—so as to cause chaos, turmoil and suffering—by coming into contact with alchemical frequencies. It is such powerful historical associations with exposing and communing with the voice of the devil that led the military to harness music as an ancient language of somatic violence. Rituals of sonic deliverance are enacted upon the demonised captives believed to be possessed by the malignant 'evils' of Islamic fundamentalism in order to break them. Military operations of acoustic exorcism call upon the recorded musical violence of the underworld to infiltrate and cleanse the living dead bodies tied down in Guantánamo's heavy-metal cells. The devil no longer has the best tunes; it has the state's tunes.

REFLECTING SILENCE

To comprehend where the cells of sonic violence emanated from, ideologically speaking, it is helpful to consider how waveformed spheres of judicial and military punishment and torture have radically changed over past centuries. In the West, twenty-first-century practices of sonic punishment have come to diametrically oppose those of the eighteenth and nineteenth centuries. In those earlier epochs, delivery from sin for the imprisoned captive was conceived of as being possible only through redemption and atonement (both concepts informed by Christian theology). These church- and state-mandated psychological imperatives were subsequently instigated and enforced through rhythmically orchestrated periods of what was understood to be the most conducive waveformed context for such self-studying contemplation: silent reflection.

Researching prisoners who were put into solitary confinement as punishment for making noise in their cells in the mid-nineteenth-century United States, Mark Smith notes that 'the control of inmates' minds and bodies via the "silent system" was an important component of discipline in antebellum Northern penitentiaries and on Southern plantations' (Smith 2003, 141). Echoing this finding, Bruce Johnson and Martin Cloonan state that 'an account of the English prison system was published by reformer John Howard in 1777 and included a number of proposals for improving the institution: "Solitude and silence are favourable to reflection and may possibly lead to repentance"' (Johnson and Cloonan 2009, 38). What is arresting here is the fact that, while the technique of alienating prisoners from social, physical and psychological relations continues to be practiced by the state (in the form of isolation chambers), the role played by the frequency-based domain has changed considerably. Whereas the psychological rationale of the eighteenth-century cell reflected the puissance of solitude and silence, Guantánamo's heavy-metal cells are haptically composed through shackled isolation, and their (dis)order is amplified by sonic force. The only quiet left in these contemporary offshore cells is in the periods of silence purposefully left between bouts of torture, giving the detainee time to foment feelings of fear and anxiety before the next session begins. Effectively it would seem that we have come full circle, to a situation where a detainee's silence leads to incarcerated sonic confinement for days or weeks at a time.

The reasons for this shift in practices and this new orchestration of acoustic excess become more obvious when it is considered how they compare to the quiet logic proposed by English philanthropist and prison reformer Howard. In eighteenth-century England, 'Howard's reflections heralded a new phase in the philosophy of state incarceration, with silence becoming the signifier and the driver of civil obedience and deference to the rule of law' (Johnson and Cloonan 2009, 38). Since no international rule of law is applicable or respected within the detainment camps of Guantánamo Bay, a new sonic logic of detention and torture has been fabricated by the US military, constructed from a different set of dynamics altogether. Here, acoustic terror—resonant with chaos and disorder—becomes symbolic of and synonymous with the state's wilful neglect and disregard for internationally agreed treaties, protocols and conventions. The intense sound of the cells thus represents the wider illegality of the detainment camp and its practices, which are as anarchic as they are excessive, as contingent as they are disturbing.

Implicit within the prison system of the eighteenth and nineteenth centuries was the behavioural directive of 'maintaining reflection by the rule of silence' (Foucault 1975, 238), bringing us to the second reason why the torturous music of Guantánamo resounds with displaced rationality. In Guantánamo there is no state engagement with the concepts of expiation or redemption. The reasons for abandoning silence and inverting the acoustic modus operandi have an illicit logic to them—a rationale that shackles sonic *doxa* and forces it into positions that contort its once accepted sensibilities. Thus, in considering Guantánamo's frequency-based confines, it would be somewhat foolish to think that 'because music represents the ultimate intoxication of life, it is carefully placed in a container of silence' (Schafer 1993, 257). Instead we are obliged to speculate upon static containers of noise and the speakers carefully placed within them—sound systems that transmit the ultimate sonic afflictions into the psychologies of Guantanamo's 'living dead'.

When deliberating upon this intense sonic pressure, the question inevitably arises as to whether it can still be called *music* or it is simply *noise*. For the sake of argument, if music can be basically defined as the organisation and ordering of sonic waveforms so as to form cultural narratives, then, conversely, 'noise takes sound *out of order*. It's chaos' (Henriques 2003, 457; emphasis original). For detainees enduring sonic torture, of course, it matters little whether it is labelled noise or music; they are battling to retain the cognitive mechanisms that would render such a question meaningful at all. It is in fact more useful to think about what the military perpetrators of torture wish this sound to be perceived as. One thing is for certain: They do not want it to be comprehended as musical, for this would defeat the psychologically strategic purpose of transmitting it in the first place.

Played at unbearably loud levels, the sonic payload distorts so that the detainee loses track of its initial compositional structure through the displacing dynamics of overload. When this excessive waveformed velocity is materialised, the sonic dominance takes him 'to the sound barrier—that is the edge of sound. Electronic amplification pushes the sonic to the limits. On one side is music, on the other noise. On one side is regulation, modulations and moderation, on the other is irregularity, unpredictability and excess' (ibid.).

Maintaining control over how and where waveforms are positioned and being able to manipulate them is the key to harnessing frequencies as weapons. Without agency over the transgressive capacity of sound, it remains unmanageable. Helping conduct these rhythms of collapse, the repetitious nature of the torture is amplified in the expectation that the detainee will slowly begin to lose all semblance of perceptual structure as he tries to disengage from his sensory context. If such a loss of definition occurs and the mind ends up perceiving only chaos, then the military will have succeeded in turning music into noise. They will have broken down those phenomena representing organisation and order and replaced them with a reverberating system that perpetually echoes the unpredictable nature of disorientation.

Distinguishing music from noise or noise from music is continually problematic because the definitions of each are in a constant state of oscillation. As French musical semiotician Jean-Jacques Nattiez tells us, 'The border between music and noise is always culturally defined—which implies that, even within a single society, this border does not always pass through the same place; in short, there is rarely a consensus' (Nattiez 1990, 48). The state uses this malleability of meaning to its advantage when communicating with the media: Employing music as torture sounds less abusive and violent than employing 'noise' as torture. We associate 'noise' with the sonic by-products of industrial processes, loud neighbours, physiological damage such as tinnitus or other acoustic phenomena that we dislike or detest. The thresholds at which we culturally agree that music becomes noise or noise becomes music are virtually impossible to locate with any precision. To conclude, if the meaning of sound constantly oscillates between clear and distinct definitions, and if a waveformed taxonomy is therefore ideologically dependent upon context, then just as music can always become noise, so noise can always become music.

Although they do not use the term *noise*, Michael Bull and Les Back directly acknowledge the capacity for sound's meaning to oscillate as it operates between polar distinctions. They write that music 'has both utopian and dystopian associations: It enables individuals to create intimate, manageable and aestheticised spaces to inhabit, but it can also become an unwanted and deafening roar threatening the body politic of the subject' (Bull and Back 2003, 1).

Attali further distinguishes the latent violent potential of noise when he declares that 'in its biological reality, noise is a source of pain. Beyond a certain limit, it becomes an immaterial weapon of death' (Attali 1985, 27). Goodman then develops Attali's notion of sonic disorder when he theorises the dissonant spatiality brought about by a postnoise condition. He writes, 'Noise, in fact, as it scrambles music's signal, destroys, for Attali, the coding regime, transforming the relationship between inside and outside and spawning a new musical order in the aftershock of its arrival' (Goodman 2009, 51).

Following their arrival in Guantánamo, detainees are destined to exist within states of sonic chaos that anticipate the neurological noise (psychological disorder) that is to come. And when this tumult happens, there are no new musical orders. In fact, there is no new music at all—just a continuing frisson of disconnected feedback to be endured, a never-ending series of aftershocks that constantly destroy the ordering capacity of the captive's mind. R. Murray Schafer anticipates this damaging potential inherent to the sonic when he reminds us that 'man has always tried to destroy his enemies with terrible noises' (Schafer 1993, 28) and then follows this with the observation that the controlled violence of noise is so dynamic and effective that it has been employed throughout recent history to strategically repress other cultures. As he explains, 'It is Europe and North America which have, in recent centuries, masterminded various schemes designed to dominate other peoples and value systems, and subjugation by Noise has played no small part in these schemes' (ibid., 77). By extension, this implies that, since they possess the capacity to bring more intense levels of abstract sonic force to bear upon another, waveforms can be made to function as means of psychological dominance. It goes without saying that the voice is a powerful conveyor of this dynamic.

VIOLENCE AND THE VOICE

> A voice can kill. A voice can destroy. A voice can be engineered to burst from a riven but resilient body.
>
> —Matthew Fuller (2005, 31)

There is an interesting ambiguity to this statement of Fuller's. Is the voice bursting from the body with intent to resist, or is it being forced to burst from the body by others who wish to expose and break it? Or could it be that there is another voice at play here, a thirded articulation of presence that reserves its right to somatic interiority?

In the Guantánamo cell there are indeed three articulations that must be considered in order to learn more about how the human voice is engineered to reveal, hide and resist. The first voice is amplified through the military's litany of commands, verbal abuse and interrogative procedures; the second voice is the detainee's, characterised by its silence as much as by its uttering of prayers, communications and expressions of pain; but the third is what the state purports and projects to be the enemy's hidden articulation—and this is the target the military seeks to expose, by drawing out the detainee's internalised or 'inner voice'.

That the majority of detainees have nothing to hide is of no concern to a military machine intent on venting state violence upon those perceived to be the enemy (irrespective of whether the laws and beliefs they 'defend' have been transgressed or not). In her seminal text *The Body in Pain*, Elaine Scarry reinforces the supposition that state violence is rarely actuated in pursuit of a direct aim. She writes, 'What masquerades as the motive for torture is a fiction. The idea that the need for information is the motive for the physical cruelty arises from the tone and form of the questioning rather than from its content: the questions, no matter how contemptuously irrelevant their content, are announced, delivered, as *though* they motivated the cruelty, *as if* the answers to them were crucial' (Scarry 1985, 28–29).

In an intricate and detailed manner, Scarry subsequently outlines the inexpressibility of pain while analysing the role played by the voices of interrogator and victim in the dynamics of torture. In one telling passage, she argues that 'through his ability to project words and sounds out into his environment, a human being inhabits, humanizes, and makes his own a space much larger than that occupied by his body alone. This space, always contracted under repressive regimes, is in torture almost wholly eliminated. The "it" in "Get it out of him" refers not just to a piece of information but to the capacity for speech itself' (ibid., 49).

Scarry is quite right to attribute such importance to the spatial extension of the self via the sonic channel of the voice. The acoustic domain facilitates this extension of the self, the other and the body, along with their claims to agency. For the detainee to have his agency drowned out and walled in by sonic phenomena subjects him to the torment of isolation and psychic despair. Shackled in a cell, he has no access to such self-extension, and meanwhile, intense waveformed pressure is directed against him. The channels usually available for distribution of oneself are backtracked and introverted so that aural trajectories start to burrow back into the psyche, scrambling the coherence of reason into feedback loops of interiority. Scarry neatly summarises these processes of spatial excess and suppressed articulation when she writes that 'ultimate domination requires that the prisoner's ground become increasingly physical and the torturer's increasingly verbal, that the prisoner become a colossal body with no voice and the torturer a colossal voice (a voice composed of two voices) with no body' (ibid., 57).

The problem with such a Manichaean portrayal of the situation is that it does not take into account the messy, distorted anatomy of the cell's sonic operation. In fact, Scarry's notion of the self-extending torturer does not go far enough in terms of explicating how the latter transgresses the intimate spatialities of the detainee. It fails to implicate the captor in the act of extension into the very being of the detainee; consequently, Scarry's analysis stops short of describing the psychologically knotted and troubled search for the third voice.

Reading Guantánamo's operations of sonic torture through Scarry's work raises a number of problematic issues, beginning with this lack of acknowledgement of the third voice. Scarry predicates her arguments and analysis on the binary constructions of the torturer and the victim, the voice and the body and extension and silence, ignoring the inevitable thirded presence within the psychological structuring of torture. Attending to the third voice in the cell implies that we question the zero-sum game that would pit the detainee's voice against the interrogator's voice and where one must be silenced if the other is amplified. In fact, the voice that is silenced is not the one the state seeks to intercept; the voice they want to hear is the detainee's internalised articulation, perceived as holding, and holding back, valuable information. The interrogators' verbal abuse, mocking tones and questions are present throughout both the sonic torture sessions and the interrogations that follow them. In response, the shouts, screams and pleading voices of the detainees have reverberated from the cells back into households around the world, through films, websites, recordings and newspaper reports. But the dynamic of the internalised voice—the third voice—is a different entity altogether, because it is so difficult to access and amplify, even when the subject has been broken.

As suggested by revelations from a number of those who endured such techniques, conceiving of sonic torture in dualistic terms is not the only problem when trying to apply Scarry's theories to the psychological and sonic conditioning of captives in Guantánamo. Scarry's assertions about the roles of body, voice and psychic pain need to be reconsidered in the light of claims made by detainees. She posits that 'the infliction of physical pain' is the most emphatic modality of violence that can be effectuated and propagated by the state upon a captive (ibid., 19) and also suggests that somatic violence is the initial encounter that kick-starts all ensuing rhythms of ritualised abuse. Yet a number of reports have specified that psychological damage caused by sonic torture was what detainees feared most, more than any physical damage inflicted upon them (Rapp 2010). Scarry's hypothesis that 'physical pain is able to obliterate psychological pain because it obliterates all psychological content, painful, pleasurable, and neutral' (Scarry 1985, 34) thus proves inadequate when considering the martial rhythms of psychological torture in Guantánamo. At no point does she address sonic torture's no-touch capacity to generate lacerating psychological and physiological pain.

Despite these problems with the ways in which Scarry theorises the sonic through the production of pain, she does hint at the importance of the acoustic domain as a sensorial dimension in which critical transformations occur—from communication to silence to internalisation, from breakdown to reconnection and the extraction of information. 'In this world of broken and severed voices, it is not surprising that the most powerful and healing moment is often that in which a human voice, though still severed, floating free, somehow reaches the person whose sole reality had become his own unthinkable isolation, his deep corporeal engulfment' (ibid., 50).

Scarry does not further develop this recognition of the crucial role played by the voice (and other waveformed phenomena) in spatialising and organising the production of pain within contemporary no-touch torture practices. Her deservedly influential text fails to address the fact that psychological forms of torture—such as intense sonic pressure—have been in use globally since the mid-1970s. It has seemingly taken the extreme noise-ridden aberrations of sites such as Guantánamo Bay and Abu Ghraib to make philosophers, reporters and musicians stop believing their eyes and listen instead; and we have to listen intently because, when the state wants us to hear the objects of pain rather than look at them, it has sought to mute the transmission of their inherent violence.

THE INSTRUMENTALITY OF SONIC PAIN

Conceptions of agency over and instrumentalisation of music are pivotal to the dynamics of how one situates oneself or is situated in the waveformed domain. This is an elementary proposition but an important one, as Johnson and Cloonan confirm when they state that 'the sense of control is central to many negative reactions to music' (Johnson and Cloonan 2009, 24). This basic insight prompts speculation about the claims that frequencies make on the psychological, spatial and physiological self. We have already ascertained music's capacity to extend the self and its aesthetic predilections, political sympathies and transgressive desires into other spatial domains, so we can also assume that these characteristics manifest themselves in the presence of others who inhabit those spaces—in which case the sonic becomes a carrier or messenger of personal information, delivering one's will into the intimate domain of another.

Indeed, 'of all the elements in the modern soundscape, music is among the most invasive, because over and above basic sonority, it projects finely discriminated markers of social difference such as taste, class, race, age and gender' (ibid., 2009, 163). A prime example of the propensity of the sonic to reveal and expose information normally regarded as private is the experience of hearing a next-door neighbour's music. We are annoyed not so much by the sound itself as by the fact that we are receiving intimate signals about another's character and temperament. We are compelled, without any choice in the matter, to perceive another's feelings and desires in sonic form. It is often this knowledge that causes anxiety and subsequently triggers strained responses, while also organising our own perception of presence within a space. Among sensory phenomena it is only music that has the capacity to assume this role of intrusive psychological harbinger and interlocutor. If we wish to extend our feelings of melancholy, frustration or joy we often turn first to music to extend such emotions that we may have difficulty fully expressing via language. Taken to its sonological conclusion, in terms of transmitting the microrhythms and layered oscillations that orchestrate the self, we may propose that music takes up where language's expressive capacities end.

In light of the above, we may agree, then, that 'relations of power (who chooses the music, and the conditions under which it is experienced) can enable any music to arouse aggressive forces' (ibid., 146). In Guantánamo, of course, the repetitious music played to the detainees was intended to pacify and rupture them rather than to make them aggressive. And yet the music played by those who tortured succeeded in instilling more aggressive and confrontational behaviours within the soldiers themselves, in turn vindicating the employment of violence in their daily routines of subjugating detainees at the whim of the US state. Power relations are obviously vital to whether sound is perceived as torture: Roman Vinokur reports that 'the reaction of human bodies to the same noise can also be very different and even anomalous (for example, an experimental study done in Russia revealed that about 7 per cent of cadets from one military engineering school that were tested performed better under high-intensity noise than in normal conditions)' (Vinokur 2004, 20).

None of the above is intended to echo Schafer's arguments against the malign effect of increasing levels of sonic detritus in the world[10]. Through explicating the acoustic dynamics of intimate spatiality, connections can be forged between apparently divergent sonic conflicts, the local noise complaint and remote Guantánamo sonic intensity coming into conversation with each other around the realisation that when 'one no longer has ownership of one's own sounds [it] is a profound and painful violation' (Johnson and Cloonan 2009, 158).

FEAR OF MUSIC

> Make people Forget, make them Believe, Silence them. In all three cases, music is a tool of power: of ritual power when it is a question of making people forget the fear of violence; of representative power when it is a question of making them believe in order and harmony; and of bureaucratic power when it is a question of silencing those who oppose it. Thus music localises and specifies power, because it marks and regiments the rare noises that cultures, in their normalisation of behaviour, see fit to authorise. Music accounts for them. It makes them audible.
>
> —Jacques Attali (1985, 19)

Throughout modern and ancient history the creation of fear has been one of these authorised uses of music. In the early part of the fourteenth century, Europeans began packing gunpowder into firearms so as to increase their military capacity to kill from a distance, but 'it was another trait of expression of gunpowder that made it a success: the loud noise it made as the explosion was actualised had a powerful effect on the enemy's morale' (DeLanda 1991, 30). Along with the anxiety associated with the archaic sonic signature of gunpowder, fear was generated by the visualisation of the weapon: 'the mental habit of recognising pain in the weapon (despite the fact that an inanimate object cannot "have pain" or any other sentient experience) is both an ancient and an enduring one' (Scarry 1985, 16).

The weapons examined here—speakers—are common everyday objects most regularly associated with the transmission of pleasure. The observation that seeing is no longer believing can in part be attributed to the rise of and increased reliance on sonic forms of communication, amplification and recording. And yet, ironically, a similar sense of evidentiary detachment is now occurring with respect to *listening*, as we are enveloped in an era of distrust of the waveform. It was through powerful speaker systems that the Ghost Army sound engineers of World War II helped sonically compose such sensorial disconnect, and it is they who haunt the deceptively resonant bodies of the Guantánamo detainees—who would not have wanted to hear or believe what sonically bombarded them (University of Michigan's Hatcher Library 2010).

In the torture cell, the inevitable nervousness of each detainee was constant, entrained and fostered to create anxiety, as internal pressure systems preempted the next torture session. Never before has the intuition that 'listening to music by oneself restores, refreshes, and heals' (Storr 1992, 122) seemed so optimistic and hopelessly utopian. We now know that such listening activities, remodulated and intensified, can also root down into neural networks, creating schizoid[11] overtones that decompose the very notion of 'oneself', but perhaps this is not so much a conclusion as a first introduction to comprehending pressured, waveformed subjectivity.

FUTILITY MUSIC

Music has often been attributed with the capacity to discover schizophrenic dimensions among the psychological foundations that support lucid, knowledgeable and rational architectures of thought (Abe, Arai and Itokawa 2017). Over the following pages, the modalities and rhythms of these dimensions, along with the potentialities of the sonic to reorganise basic principles of cognition, will be considered in relation to acoustic-torture practices. A brief explanation of how the 'schizophrenic' is situated and conceived of will be informative here. Bypassing traditional psychoanalytic theory, Fredric Jameson succinctly suggests that 'with the breakdown of the signifying chain, . . . the schizophrenic is reduced to an experience of pure material Signifiers or in other words of a series of pure and unrelated presents in time' (Jameson 1991, 72).

This notion of disconnection—from oneself, the other and the world at large—will be adopted as an underpinning for the following analysis, which explores a number of diverse perspectives explicating the relations between music and the orchestration of schizoid practices, psychologies and spatialities.

Nineteenth-century politician and soldier John A. Logan declared, 'Music is the medicine of the mind' (Logan as quoted in Storr 1992). In the twenty-first century, in the hands of the military-entertainment complex and through its speaker systems, this 'medicine' has been transmogrified into a poison. Thanks to contemporary mobile-playback and amplification technologies, this process of malign conversion has become insidiously straightforward: select a track, play at high volume, repeat as long as is deemed necessary. Indeed, the power of such a technique lies in its simplicity. It is a process determined by the illogical excesses of sonic overload, which have the potential 'to circumvent normal rational process' (Henriques 2003, 457), and is a phenomenon to which every man, woman and child capable of hearing is vulnerable. In trying to expand our appreciation of music's capacity to endanger basic psychological functionality, it must be assumed that sound is, to some degree, neurologically hardwired to assist in engineering our cognitive behaviours. As psychologist Leda Cosmides and anthropologist John Tooby once wrote, 'Music's function in the developing child is to help prepare its mind for a number of complex cognitive and social activities, exercising the brain so that it will be ready for the demands placed on it by language and social interaction' (Cosmides and Tooby quoted in Levitin 2006, 262).

It is plausible to suppose that a faculty that assists in the construction of rationality could subsequently be subverted, reversing that constructive process and compromising the mind's ability to create connections.

As established earlier in this chapter, when under the duress of sonic pressure, detainees at Guantánamo were impelled to escape psychological and physiological abuse through internalised processes that eschewed connections and interactions with their captors. In this attempt to mentally remove themselves or to shelter their psyche (symbolised by the inner voice) from their immediate painful situation, a numbness to sensorially induced affect had to be achieved. This would have resulted in a perceived emotional retreat from said context, with outward similarities to some of the traits of hebephrenia[12]. For Theodor Adorno, 'Hebephrenia is finally revealed from a musical perspective to be what the psychiatrists claim it to be. The "indifference towards the world" results in the removal of all emotional affect from the non-ego and, further, in narcissistic indifference towards the lot of man. This indifference is celebrated aesthetically as the meaning of this lot' (Adorno 1973, 129).

That in Guantánamo temporary attainment of such a psychological state of disconnect had become a necessary survival tactic—or, more accurately, an expedient 'sanity tactic'—is a sombre musical irony that would not have been lost on the Frankfurt School theorist.

As investigated by Gordon Thomas (1989) throughout the past century and predating these sonic techniques for inducing schizoid states, a number of medical experiments were conducted using sound, sometimes alongside drugs and electroconvulsive therapy. Endeavouring to understand both how to correct and how to induce psychological states such as hebephrenia, reality distortion and psychomotor poverty[13] (understood to be the three manifestations composing schizophrenia), the CIA employed Donald Ewen Cameron to conduct experiments for Project MKUltra at the Allan Memorial Institute of McGill University in Montreal between 1957 and 1964. Cameron's most remarked-upon procedure was called *psychic driving*, which involved 'depatterning' subjects—or putting 'patients' in drug-induced comas for long periods of time (up to three months) while playing sonic content to them repeatedly via tape machines under their pillows. The sound would vary, at times consisting of noise, at others positive or negative messages such as, 'Madeleine, you let your mother and father treat you as a child all through your single life', followed by 'You mean to get well' (Rejali 2007, 370).

The parallels between Project MKUltra methods and the repetitive sonic-torture techniques employed by the US military in Guantánamo are of consequence. Importantly, however, the ambitions of the CIA and the US military diverge at resonant points in their practices. Whereas the CIA was interested in erasing minds so that they would become akin to blank tapes onto which new programs of behaviour could be recorded, the US military was more interested in breaking down the cognitive mechanisms of the outer voice so that they could extract and amplify information from the unprotected internal voice. Despite striving to rerecord memories and compose fictitious histories, however, 'Cameron was able to destroy minds but not rebuild them' (ibid., 371). The US military, meanwhile, was able to decompose cognitive behaviours but, conversely, had no interest in reorchestrating them.

For the state, it appears that the most telling difficulties of conducting no-touch torture in Guantánamo were associated with their apparent lack of measuring techniques—that is, systems that would allow the US military to gauge the effectiveness of the sonic pressure they were applying. Remarking upon one loquacious source's comments, Cusick leaves us in no doubt as to the severity of these martial techniques and their worst intentions to induce palpable psychological afflictions. She writes, 'This modern system aims to combine "sensory disorientation" . . . so as to cause a prisoner's very "identity to disintegrate". Whether that disintegration takes the form of induced regression (to infantile behavior) or induced schizophrenia, the experimental data showed this "modern system of torture" to be much more efficient than beatings or starvation, producing psychological disintegration in a matter of days, rather than weeks or months. And, as one CIA researcher noted, it was hard to document' (Cusick 2006).

Whereas quantifying and naming psychological damage was challenging for the state, the official classification of the sonic tool utilised in these 'hard-to-document' operations appears to have been simpler, as revealed in Bayoumi's 'Disco Inferno' article: 'The Pentagon's Schmidt investigation identifies it as "futility music"—that is to say, screamingly loud and deliberately Western music that will, per the Army field manual, "highlight the futility of the detainee's situation"' (Bayoumi 2005).

Militarily speaking, the purposes of synthesising such futility and despair are many, testifying not only to the brutal nature of the state but also to its ability to provoke unnatural acts of self-mutilation in its captives as it 'produces listeners who are deprived of reason, becoming mimetic puppets of senseless sounds and excessive, destructive emotions' (Naddaff 2009). While this description in Naddaff's conference paper 'No Blood, No Foul: Listening to Music at Guantánamo Bay' may elucidate the effects of sonic torture, it is in actual fact extracted from a paragraph explaining the reasoning behind

Plato's bid to censor the practice of *poiesis*—a mimetic art form that, she explains, consists of both words and music. Plato's fears, it seems, find an echo within contemporary sonic environments. Afraid of the saturnalian extremes that music could provoke in the subject, Plato attempted to construct a legal and cultural gate to effectively cage the beast within—a sonic restraining order that would choke the excessive nature of wo/man, should it ever try to break out and exceed its somatic limits.

Within the realms of conflict and its constitutive violences (including no-touch torture), the capacity to be naturally brutal and thus monstrous is a prerequisite for operational success. As the West diligently characterises the Islamic fundamentalist subject as a monstrous inversion of all that is reasonable, comprehensible and civilised, so the Occident concomitantly demands that its own citizens and military become equivalently aberrant in their methods of attack and retaliation. In such circumstances, Virilio's summary of the practices that coordinate preternatural violence between adversaries rings truer than ever. He writes, 'A phenomenon of mimetic training, a reciprocal apprenticeship of the horror of the other, war is always a school, a university of shared terror where, bit by bit, we become like our enemy by dint of opposing them' (Virilio 2002, 57).

In the heavy-metal music cells, diverging manifestations of the monstrous come face-to-face. Targeted and objectified by the state, the detainee is sonically projected as the monstrous embodiment of psychological chaos, irrationality and waveformed rage. The torturer, meanwhile, must be transformed into an organised and controlled performer of the state's monstrous capacity for vengeance and violence. He will carry out 'a regulated practice obeying a well-defined procedure' (Foucault 1975, 40) and in the process will become a master of sonic excess.

The dynamic of this theorisation is an essentially Foucauldian one, as further supported by Foucault's claim that 'torture is a technique . . . not an extreme expression of lawless rage' (ibid., 33). Yet what has become apparent from the myriad reports, interviews and documentaries on Guantánamo is that an uncontrolled violence did manifest itself within the deranged everyday existence of the camps, through cases of sexual abuse and the savage random beatings frequently meted out—and, of course, through practices of sonic torture (employing genres of music that themselves document alienated subjects engaging in inhuman and monstrous acts). The proposition that an uncontrolled and barbarous martial nature was not only tangibly present in the camp but also had been sanctioned by the state is further substantiated by the soldiers themselves, who told Jonathan Pieslak, 'War is people having to step outside of themselves. It is you having to become what I consider to be a monster' and 'You've got to become inhuman to do inhuman things'

(Pieslak 2009, 161, 162). As Pieslak subsequently notes, 'Psychologists Robert W. Rieber and Robert J. Kelly observe that the circumstances of war can cause combatants to develop inhuman or dehumanizing views: "Self and object-directed dehumanisation are inevitably heightened in situations where the threat of combat is present". In these instances, the music could be said to have a transformative power that removes the humanity element from human identity. Music becomes a means of dehumanizing an adversary or oneself' (ibid., 163).

While here Pieslak is speaking more specifically about soldiers preparing for battle in Iraq, his analysis is pertinent to the way in which music is used in Guantánamo to martially transubstantiate the subject, composing a form devoid of the characteristics one would regularly attribute to a living human, such as consciousness, cognition and compassion. In this sense, the dehumanising results of sonic torture echo the legal status of the detainees, disembodying Foucault's torturing protagonists so that they also come to represent the monstrous and inhuman in the form of the living dead. If Adorno is correct in asserting that 'modern music sees absolute oblivion as its goal' (Adorno 1973, 99), then the music of Guantánamo soundtracks a 'haunted modernism', a tormented existence in which the wandering waveforms of music have returned to feed back on their own material producers. By transforming those who were once alive into the undead, and by torturing those who already represent the living dead, the music of Guantánamo Bay transmits to us its goal: to orchestrate the absolute oblivion of everything that embodies the human.

THE REPETITIVE CELL

Western strategies of musical and noise-based repetition have been impelled to their political, psychological, physiological and sonic conclusions by the torture practices employed in Guantánamo Bay. In other words, this is the end of the line for repetition as a technique. Whatever follows in terms of organisational procedure and structure will rely on different waveformed modalities to achieve its goals. Repetitive behaviours have always been integral to the disciplined practices of institutions such as military organisations. To begin with, they rely on steadfast rhythms to organise the procedures, training and collective identity of their personnel. In the sixteenth century, 'drill produced entrainment. That is, soldiers became "oscillating entities", repeating the steps of a cycle over and over, and this created a strong bond among them, the unit of cohesion that alone guaranteed the continuity of command needed in a war machine' (DeLanda 1991, 58).

For Manuel DeLanda, this entrainment through repetitive commands and actions signalled the first time that a methodological approach to a soldier's psychological and physiological preparation had been institutionalised (the Dutch army of Prince Maurice of Orange being the first military to practice drilling sessions in their daily routines). However, in locations such as prisons, repetition serves a different mandate—one that organises detention, pain and discipline in order to prevent legal transgressions from ever occurring again. Thus, for Foucault, the prison rationale demands that 'one must calculate a penalty in terms not of the crime, but of its possible repetition. One must take into account not the past offence, but the future disorder. Things must be so arranged that the malefactor can have neither any desire to repeat his offence, nor any possibility of having imitators' (Foucault 1975, 93). The bearer of divergent significations in the prison and military systems, repetition is versatile, nonpartisan and central to the ideological constructs of both yet bound to neither.

From a more-enveloping perspective, Attali's conceptualisations of cultural repetition within twentieth-century Western capitalism led him to claim that 'repetition today does indeed seem to be succeeding in trapping death in the object and accumulating its recording. . . . Everything in our societies today points to the emplacement of the process of repetition' (Attali 1985, 126–27). He identifies the imperative born of Thomas Edison's 1877 invention of the phonograph (a technology capable of both recording and reproducing sound)—that of dropping 'the end' into a feedback loop that never has to finish. Music never again has to encounter sonic death. After the phonograph, and even more so in the digital era, it is produced as an object that can be stockpiled, played back and distributed ad infinitum without any material or acoustic degradation. Musical genres have been born and have evolved into compositional modes that embrace repetition as the heartbeat of their sonic construction. Add to this forms of dissemination such as mainstream radio that transmit the same tracks every hour, day and night, and it becomes apparent that we are audience to strategies of sonic repetition on an everyday basis. Ever since the moment we attained the power to play back a recording, we have gradually increased the speed and intensity of those repetitions—up to the point in Guantánamo where music realises its innate potential to be played relentlessly, without silences, without breaks between transmissions.

In the torture cell, the spaces in between waveforms are collapsed into a stream of noise that has no time for any redemptive interstices. This reverberating channel is the final acoustic solution—a sonic manifestation of violence that becomes audible via the annihilation of silences and is amplified by the creation of pain through pure excess. The historical lineage of this channel weaves in and out of civilian and military histories that have amalgamated into the present form of the military-entertainment complex, an inevitable act of acoustically forced copulation, given that sonic 'violence is no longer limited to the battlefield or the concert hall but pervades all of society' (Attali 1985, 36). With the soundtracks of everyday living being orchestrated by this military-civilian coupling, distinctions between the cell and the home have become not so much ideological, political or economic as temporal. They are distinctions born of a chronological taxonomy for which the signifiers of difference between perceived sanity and breakdown reside in the units of time between repeats. As such, the assimilation of music by the military-entertainment complex into a direct weapon has forever changed the political, aesthetic and cultural standing of sonic expression. Attali foresaw this radical shift in the vital agency of organised sound when he declared that 'music, exploring in this way the totality of sound matter, has today followed this its path to the end, to the point of the suicide of form' (ibid., 83). With this statement in mind, it is not overly dramatic to propose that music can never be thought about in the same way again. The final nail in the coffin of music's period of innocence has been hammered home by this military assimilation; the only question as to the demise of its identity as nothing but a form of cultural expression is whether this end is to be documented as assisted self-destruction or murder.

If music is witnessing its own demise, then it is simultaneously predicting the incremental dissolution of other cultural systems of organisation, giving credence to the idea that 'music today is in many respects the monotonous herald of death' (ibid., 125). The detainee can be placed in the crosshairs of Attali's prophecy: the living dead in Guantánamo enduring endless sonic invocations of their locked-down situation. They are offered more of the same or ultimate cessation by a military that champions the postmodern fracturing and death of the subject. Alternatively, the statement could be translated as anticipating a more-inclusive kind of death, that of an entire culture: the US military's sonic-torture policy deciphered as a virulent cultural offensive against Muslim societies. Having infected Muslim detainees with contagious sonic cells, the United States will subsequently send them back to their homelands where their symptomatic experiences will signify the ensuing threat of future propagation. Put another way, Guantánamo's music not only represents an assault on individuals, it also symbolically announces the intentions of the US military-entertainment complex to ideologically propagate and transmit their culture into Muslim iPods, living rooms, shopping malls, public-transportation systems and streets.

With powerful global-communication networks, production facilities and distribution systems at their disposal, this implicit threat on the part of the United States is anything but idle. The culturally imperialistic timbres embedded within the Western voices of the camp guard, along with the articulations of the rappers and rock singers whose voices are employed to torture, come to represent the tonal currency of a moral chaos for the Muslim subject and state, the individuated narratives of Western excess predicting the future degeneration of a collective religious ordering. As Foucault rightly points out, 'In the "excesses" of torture, a whole economy of power is invested' (Foucault 1975, 35). The power implicit within Guantánamo's musical torture resides in the threat of contagion and in the potential for such excessive musical and sexual practices, overflowing with seductive intent, to become irreconcilably embedded within the Muslim's naturalised surroundings and systems of living. The Occidental weaponising of culture through repetitious music, therefore, comes to signify the virtuoso capacity of the viral to endlessly oscillate and replicate. It also makes clear music's cultural status as the most infectious and readily transmittable armament, given its capacity to both channel excess and amplify its instrumentality at the same time.

In *The Parallax View*, Slavoj Žižek takes a similar perspective on the imperialistic torture practices of the United States, albeit one that identifies a crucial flaw in its logic. He writes, 'Abu Ghraib was not simply a case of American arrogance toward a Third World people: in being submitted to humiliating tortures, the Iraqi prisoners were in effect *initiated into American culture*; they got the taste of its obscene underside which forms the necessary supplement to the public values of personal dignity, democracy, and freedom' (Žižek 2006, 370; emphasis original).

Žižek's premise that the ritualistic nature of the torture in Guantánamo and Abu Ghraib represents some form of initiation ceremony is, however, misleading. The initiation or hazing rituals of 'secret' societies and clubs such as Yale University's Skull and Bones (whose alumni include former presidents George H. W. Bush and George W. Bush, as well as US senators such as John Kerry), military inner-circle cabals and criminal organisations (such as the triads, Cosa Nostra or Yakuza) are undertaken by subjects with the tacit understanding that acceptance will be gained upon successful completion of the ritual. For those who undergo initiation, there is a gratification in being in receipt of ritualistic suffering when it is understood that it must cease and the victim will eventually be sanctioned to carry out similar tasks in the future, their victimhood shed and passed on. In Guantánamo, there were and are no options for such role reversal, either within the torture cell or when (if) the detainee leaves.

FOUR WALLS OF SOUND

Composing virasonic sorties through sprawling global communication networks requires a comprehension of rates of aesthetic infection, levels of informatic feedback loops and latitudes of asymmetric contagion. The engineering of resonant threat within the torture cells, however, required a different mode of waveformed architecture altogether in order to effectuate the annexation and contamination of the subject. The essential acoustic construction method for such undertakings is elemental in its structure, robust and easy to 'bring down' if required: it is the *wall of sound*. Mick Brown (2007) explores the history of this term derived from a recording technique developed in the 1960s by Phil Spector (in LA's Gold Star Studios) for producing rock and pop music. The basic premise of its sonic construction is to have multiple performers playing the same part in unison and to record them in an echo chamber, giving the final layered recording an impenetrable, reverberating acoustic dynamic. In a parallel martial manoeuvre, the torture cells of Guantánamo became analogous to the studios responsible for the construction of such a sonic operation.

The architectural nature of the term *wall of sound* reveals much about the cultural propensity to construct material narratives around the hapticity of waveforms. The wall of sound is an architectonic formulation predicated on the absolute physicality of frequencies and their ability to organise and construct a signature spatial experience. Waveforms can be conceived of as having the ability to psychologically construct emotions and expeditiously fabricate dwellings in which our fears, desires and aggressions can temporarily take up residence. Thought of in another way, the wall of sound falls instantly, and we are compelled to appraise the mobile nature of waveforms and their capacity to operate as *transarchitectures*—as nomadic cultural signifiers of presence and habitat that are there one moment, gone the next. It is this latter functionality that motivates the military to utilise such ephemeral practices, using walls of sound to overlay and reinscribe the limits imposed upon the sensorium by the physicality of the concrete walls. *Sonarchitecture*.

Such a proposition exposes the shift that occurred in self-orienting techniques between the nineteenth-century cells analysed by Foucault in *Discipline and Punish* and the torture rooms of Guantánamo. In the former construction, 'the walls are the punishment of the crime; the cell confronts the convict with himself; he is forced to listen to his conscience' (Foucault 1975, 239). In Guantánamo, having already established that there was no interest in eliciting self-reflection within the detainees, what we see instead is an authoritarian commitment to developing environments within which the self echoes and reverberates. The ideological shift proposed earlier was manifested in the cell by the construction of a sonic dominance that did not so much confront the detainee with himself as invert his notion of self. The negation of his physical self-extension beyond the cell was amplified by the silencing of his sonic self-extension, forcing him to interiorise his presence. Sonic walls are thus the most malevolently effective structures of power, as they can be dropped and raised at will, meaning that release from a sonarchitecture can be promised at any moment in time, only to be rescinded the next.

THE EXCESSIVE REALITY OF WAVEFORMED PRESSURE

While the massive noise of Desert Storm's incoming carnage would have been bone-shakingly close for the Iraqi people in 1991, the technological prowess of the US military allowed them to mount a logistical destruction that was largely conducted from a distance, with the majority of damage to Iraqi military and civilian infrastructures coming by way of more than one hundred thousand aerial sorties flown by coalition forces at the start of the war. The violent aftermath, however, starting with the second Persian Gulf war in March 2003, left Jean Baudrillard's (1991) claims that such conflicts are now virtual 'Nintendo Wars' looking inaccurate and territorially myopic. The long-term reality of the Iraqi invasion has consisted of anything but digitised bodies, next-level politics and CGI theatres of operations; a more exacting analysis of its actuality informs us of its noisy, close-at-hand, bloody, devastating and dirty travails.

As in the Guantánamo cells, the sonic strategies employed in Iraq (using similar types of music) were, upon first hearing, surprisingly basic in their composition, but this is where parallels between the sonic cell and the sonic battlefield end. Conducting acoustic reconnaissance on territories outside the torture cells reveals the ways in which the spaces that surround them were also politically, socially and geographically contoured by sound during conflict. There are many germane examples of waveformed methods being utilised to extend presence and to define cultural and spatial limits in conflict zones: the US military regularly travelled the streets of Iraqi cities in Humvees playing loud Western music. In one instance, citizens in Fallujah decided to acoustically engage with their sonic nemeses, obviously frustrated and offended that their sonic terrain was being dominated by music that did not echo their beliefs or values. Turning up the volume of several mosques' speakers (contained in the minarets of each building), they responded by amplifying their *nasheeds*[14], in so doing engaging in a waveformed battle of cultural and religious extension, paranoia and acoustic intimidation. While the notion of 'battle sounds' is not a new one, this modality in which music becomes the sole weapon certainly is. In the past, music has been utilised in conflict to encourage physical aggression on the battlefield, but cases such as that of the Fallujah sound clash expose a new battle dynamic—one in which the armaments of assault consist solely of electrically amplified speaker systems, track selection and the potential to deploy sound pressure. In a somewhat inevitable turn, the asymmetric techniques of conflict engagement and geographically abstract weapons systems have been spurned in favour of an 'on-the-ground' base reality—an unspoken rule saying that whoever shows up with the most powerful sound system owns the space (until a louder one comes along).

This example of speaker-system battles in Iraq ends any possible circumspection about music's potential as a direct weapon. The way in which sound is harnessed to denote range, to inscribe the circumference of engagement and to determine the nature of presence is still contentious, however. Assigning sonic affect to the negotiations of distance, Virilio declares that 'the battlefield is the place where social intercourse breaks off, where political rapprochement fails, making way for the inculcation of terror. The panoply of acts of war thus always tend to be organised at a distance, or, rather, to organise distances. Orders, in fact speech of any kind, are transmitted by long-range instruments which, in any case, are often inaudible among combatants' screams, the clash of arms, and, later, the various explosions and detonations' (Virilio 1994, 6). While there may have been no clashing of arms in the torture cells, there were screams and articulations of pain heard only faintly (if at all) by the soldiers who carried out their painful sonic practices via an organised geometry of distances that set them apart from their victims.

Historically established forms of torture, such as electrocution, the hammering of metal wedges under fingernails and beating, require the immediate presence of the torturer; but more than this, they engage the torturer in a direct and intimate physiological, psychological and spatial relationship with the captive under assault. Within the dynamics of no-touch torture, the perpetration of sonic violence necessitates that the person who tortures must keep a viable distance from their own weaponised amplifications so that they do not themselves become vulnerable to the effects of acoustic torment. Thus the torturer will delegate pain from a safe distance from outside of the room. The torturer will watch and listen in relative comfort and with little requirement for personal interaction (negating any feelings of compassion that might arise from relationships being unexpectedly formed during interrogation) while conducting logistically precise abuse. No-touch torture activates this new set of violent behaviours, spatial reasoning and impersonal relationships and in doing so converts distances into a new set of sonospatial channels between the detainee and those inflicting pain upon the detainee.

In this chapter the sonic spatiality of the cell has been explored at length, but it is also useful to think more about the immediate acoustic architecture that surrounds it and to investigate how soldiers inhabited the 'safe space' from which they took notes and observed. If 'sonic dominance helps to generate a particular sense of *place* rather than a general abstract idea of space', one that is 'unique, immediate and the place of tradition and ritual performance' (O'Dwyer 2009, 101; and O'Dwyer quoting Henriques 2003, 459), then we may construct a taxonomy of resonantly concentric spatiality. This is a frequency-based system of classification that questions whether the space immediately outside the cell was rendered abstract because it was shut off from the sonic delineation of place, whether such exteriorised spatialities were denoted by hushed reverence, jocular tones or by a scientific documenting of the ensuing violence and whether the cell or its surrounding rooms deserve the ultimate spatial degradation of being classified as a nonplace (Augé 1995, 6).

It would seem that, for Henriques, the torture cell could not be described as such, since 'acoustic space and sonic place are the antithesis to the typically postmodern "non-places" of airports, shopping malls, high streets and ATM aprons that Marc Augé discusses. Those are generic, abstract and empty spaces. Sonic spaces are by contrast specific, particular and fully impregnated with the living tradition of the moment. Each has a certain definite haecceity or "thisness" about it' (Henriques 2003, 459).

According to Augé, one of the exemplary characteristics of a nonplace is its transient functionality, which allows subjects to traverse it without friction and as quickly as possible. A second characteristic of the nonplace is its indifference to presence, inasmuch as the subject who travels through it will form no emotional attachments to it. Augé cites motorways, airports and hotel rooms as prime examples of nonplaces, which he notes are becoming ever more symptomatic of supermodernism—an epoch beyond modernism that (from Augé's perspective) is intensely committed to technology, travel networks and the transferral of information (Augé 1995). In terms of identifying the nonspatial status of the sonic torture cell, it is difficult to imagine that either of these definitions does not apply to the detainee's desired relationship with it. The timing of the transfer through the cells, however, presents us with questions regarding definition since, in terms of presence, Guantánamo time is counted in years, rather than the minutes, hours or days that usually apply to Augé's conception of passage through the nonplace.

Sonically, the walls of sound constructing the torture cells bestowed them with the 'thisness' to which Henriques alludes. Yet such notions of haecceity would have to be shut out by the detainee attempting to negate the excessive reality of the waveformed pressures he was at the centre of. Such dichotomies expose the nexus of contradictions that construct the identity of the sonic torture cell; for it is at once the quintessential form of dystopian spatial trajectory—the nonplace toward which all other nonplaces theoretically lead—and the archetypal sonic space, defined by its walls of sound. It is a construct beholden to an overflowing of humanity (when the detainee's interiors are literally pushed or beaten out of him) as well as one of the most inhuman environments (marking it out as an architecture that defines the borders of sanity). So extreme is this mapping of the subject that, in the moment before its breaking, we find ourselves at the periphery of humanity, negotiating a sonic abyss of excessive reality. In view of the torture cell's inordinate wealth of characteristics, it is implausible to come to a definitive conclusion as to its spatial status. As its position reverberates between the strategies of the military and entertainment complexes and its status oscillates between space and nonspace, noise and silence, human and inhumane, it confers upon us a responsibility of sorts—namely, that of rigorously questioning and identifying its 'living' conditions on an ongoing basis. Only when we engage with this process will we be able to perceive the torture cell's capacity to compress everything that we are and everything that we don't want to be into a 6.5- by 8-foot space.

CULTURAL COMPRESSION

It has been argued that the military-entertainment complex has progressively compressed the architectural frameworks in which it expresses its waveformed instrumentality, the structure of the cell being the apotheosis of its sound-pressure motivations. Scarry comments upon this compression in a broader context, designating it as a technique that aims to erase the detainee's sense of the world at large: 'While the room is a magnification of the body, it is simultaneously a miniaturization of the world, of civilization' (Scarry 1985, 38). She further explicates how the architectural dynamics of the cell become part of the persecuting arsenal, an element that, in particular, is crucial when inflicting sonic pain. She writes, 'The torture room is not just the setting in which the torture occurs; it is not just the space that happens to house the various instruments used for beating and burning and producing electric shock. It is itself literally converted into another weapon, into an agent of pain. All aspects of the basic structure—walls, ceiling, windows, doors—undergo this conversion' (ibid., 40).

This proposition that architectural structures are remodulated into armaments is a significant one, as it supports the hypothesis that waveformed and architectural strategies are harmoniously composed in order to privilege military-entertainment instrumentality. Even at Waco, where the architectural structure of the compound was not (as in the factory and the cell) purpose-built for such practices, the architectonic strategies orchestrated by the state to contain it were woven around sonic amplification and recording techniques, constructing a cocoon of deprivation. Here again, the strategic time signatures of the architectural landscape and the sonic domain were compounded by each other's rhythms so that they might dominate the physical, psychological and affective resonances of agency oscillating between the two. Returning to and extending Scarry's initial premise, the cell is not only representative of such composite relations between the transience of waveforms and the stasis of architecture, it is in fact the ultimate symbol of their fusion. It is the distillation of military-entertainment spatiality—a dissecting room of sonarchitectural affect.

Foucault's hypothesis that 'even if the compartments it assigns become purely ideal, the disciplinary space is always, basically, cellular' (Foucault 1975, 143) is of particular relevance here. He develops this argument by tracing the history of such forms of architectural organisation back to the monk's cell and to the religious exercising of bodily limitations and self-restraint. Rather than conceive of the cell as a terminal realisation of architectural compression (as Scarry does), Foucault identifies it as an '*enclosure*, the specification of a place heterogeneous to all others and closed in upon itself. It is the protected place of disciplinary monotony' (141; emphasis original). While it is still true that 'discipline organises an analytical space' (143), in contemporary Western culture discipline also organises a politics of displacement, an ideological mechanism that purposely transfers our attention from the cell to the club, from the theatre of operations to the movie theatre and back again at will. The identity of the sonic cell can no longer be protected or isolated because it is in a state of constant, networked transferral between military and leisure environments; it is in this communication loop that the Guantánamo detainment camp realises its defining architectural modality: that of nomadic violence.

As preparatory behaviours for conflict are installed within the domestic regimen, Virilio's prophecies pertaining to the rhythms of exchange between the military and the municipal find a new supplement in the architectonic parallels between the spaces of festival, performance and dance and the space of the torture cell. Although being shackled into a space and dancing in a space are worlds apart, in terms of inhabiting space these environments are being continuously and irreconcilably drawn closer together. Thus, for Scarry 'it is not accidental that in the torturers' idiom the room in which the brutality occurs was called the "production room" in the Philippines, the "cinema room" in South Vietnam, and the "blue lit stage" in Chile: built on these repeated acts of display and having as its purpose the production of a fantastic illusion of power, torture is a grotesque piece of compensatory drama' (Scarry 1985, 28). Such macabre traditions of characterising torture rooms as places of cultural expression have been propagated in Iraq and Guantánamo, as evidenced by Haitham al-Mallah's recollections of being sonically tortured in the 'disco' of Mosul (Bayoumi 2005).

The naming of torture cells after cultural architectures is a significant designatory act that elides historical boundaries between civilian and military spaces. The military-entertainment complex routes content from military zones into municipal zones of leisure, locating the civilian spaces wherein their violent histories will be recounted and reenacted in the future. For it is in the cinema and the theatre that the majority of future generations will come to witness, secondhand, the atrocities of torture—violence that will be remembered for where it happened as much as why it happened or who was responsible for it. This brings us to the other logic inherent to this architectural framing and naming of pressure and violence. The rationale of alluding to compressed spaces of pain as expansive places of pleasure is one that seeks to humanise the actions occurring within these rooms—activities reliant on the capacity of the soldier to extend themselves into the realm of the inhuman. As the military knows full well, it is difficult to culturally speak about that which we consider inhuman, let alone to name it. As such, the soldiers who acted out their scripted orders will have played anonymous roles, whereas the named torture rooms, over time, will take centre stage, themselves becoming the named and remembered protagonists of their aberrant productions. Via a convenient transmutation of memorial presence and guilt, the sounds, actions and personal identities of military personnel become subsumed by the architecturally distinguished characteristics of remembered violence. For future entertainment productions, this makes for better viewing and easier listening.

WORN-OUT PHANTOM LIMBS

To deliberate further upon the mediatised nature of the Guantánamo detainment camp, we are not required to project speculative futures, for this grotesquely martialled theatre has provided us with copious documentation of its actual machinations. Since 2004, new sound recordings, videos, photographs, books, interviews and reports pertaining to the abuses have appeared on a daily basis. Indeed, the regular arrival of such information suggests a state complicity in promoting this documentary wave in order to ensure that the clandestine threat of the detainment camp is extended to a global audience. Guantánamo has realised its potential as an all-encompassing symbol of immanent danger, one that is pan-national in scope. Symbolising either the nightmare scenario of globalisation or its logical conclusion, the camp's detainees have inadvertently become the United States' poster boys of systematic transfer. Commenting on the increasing rates of transience within global systems, Saskia Sassen writes, 'It is true that throughout history people have moved and through these movements constituted places. But today the articulation of territory and people is being constituted in a radically different way at least in one regard, and that is the speed with which that articulation can change' (Sassen 1998, xxxii).

The articulated movement of bodies towards Guantánamo expresses this new dynamic of accelerated transfer perfectly. To begin with, the captives' staccato movements are fast and silent as, handcuffed, hooded and earplugged, they are hurriedly and frictionlessly transported across political, geographic and economic borders to the place of incarceration. Once inside the camp, however, they are enveloped in stasis, their lack of mobility exposing the slow-motion effects of both legal and sensory entropy. Caught in a system of sustained decay, the detainees are forced to endure the compounded deterioration of their will, of waveformed political coherency and (most importantly) of perceptibility.

As Guantánamo experimented with our perceptual rationality and the capacity of our cultural attention spans, it also signalled a new waveformed direction to be taken by the military-entertainment complex. Modulating its presence between the heard and the unheard, the seen and the unseen, the named and the unnamed, it is testing our resolve to identify its procedures, to track its movements, to preserve our collective resolve to care. It is the testing ground for a new set of martial-civilian dynamics. The techniques of perceptual overload embodied in sonic torture have reached their sensory terminus, the rates of repetition and volume levels having been increased to their maximum. The advantages of operating within the realms of the perceptible have been trialled, analysed, manipulated and exhausted. It is now time for us to track the military-entertainment complex as it invests its energies into movements outside of perceptual theatres, into uncharted zones, where the maxim 'out of sound, out of mind' comes to represent an entirely new way of thinking about waveformed affect, cognition and spatiality.

While this shift into the as-yet-perceived amplifies the military's desire to inhabit and shape the silent and the invisible, it also signals an intent to mutate and transform the perceptual apparatus of the body—to extend the range of stimuli within which it can operate. This reconfiguration of the somatic will not just affect the audible and the visible. Over time, all five senses, along with those that we have only recently started to name and to comprehend—such as nociception (pain), equilibrioception (balance), proprioception and kinaesthesia (joint motion and acceleration)—will inevitably be, in technomilitary terminology, 'upgraded'. For the present, however, audible and visible channels feeding into Guantánamo Bay are dormant, and there are now only the impending departures of the detainees to be realised. The military-entertainment complex, meanwhile, will make tracks into the remote realms of the sensorium. For it is the fields of perception that will become the new battlegrounds of the twenty-first century and the cartographies of unsound into which we now inexorably move, with augmented phantom ears, eyes and organs extending the theoretical and functional capacity of our worn-out phantom limbs.

NOTES

1. Originally consisting of three camps, Camp Iguana, Camp Delta (including Camp Echo) and Camp X-Ray, Guantánamo Bay (also referred to as Gitmo) is a US military installation in Cuba. The codes of conduct practiced with regard to its detainees are able to transgress internal US codes and mechanisms of due process since the US Justice Department declared the camp to be outside the United States' legal jurisdiction. During the period of its (fully functional) operation, there were numerous vociferous objections and complaints about the treatment of its inmates from human-rights groups (such as Amnesty International and Human Rights Watch), political leaders, globally connected groups of protestors (such as the Society for Ethnomusicology) and ex-prisoners. These remonstrations are often based on reports and direct observation of and from those held in the facility and accounts (by reporters such as Julie Hyland) of brutal violations of human rights by sonic torture, sleep deprivation, prolonged constraint (being held in stress positions over long periods of time), sexual degradation, forced drugging and religious persecution.

2. A set of torture practices also deceptively referred to as the application of 'moderate physical pressure', *torture lite* methods employed by the US military to torture detainees in Guantánamo Bay included musical torture, psychological torture, sensory deprivation, starvation and thirst, sleep deprivation, waterboarding, forced standing, sexual abuse and humiliation.

3. A campaign initiated by the human-rights group Reprieve, consisting of the release of information such as a list of the top ten songs employed by the US military to torture Guantánamo detainees.

4. Rasul v. Rumsfeld is a current suit brought collectively by Jamal Udeen Al-Harith and the 'Tipton Three' against former US Secretary of Defense Donald Rumsfeld. The claimants are requesting financial compensation from Rumsfeld, who condoned the use of illegal interrogation practices upon them during their wrongful detention in Guantánamo Bay.

5. The United Kingdom, along with the United States, agreed to abide by the updated third convention (relative to the treatment of prisoners of war) and the fourth (relative to the protection of civilian persons in time of war) in 1949, which legally dictate the humane treatment of civilians during wartime and the procedures to be abided by in relation to caring for prisoners of war. In relation to the war in Iraq, article 17 of the third convention decrees: 'No physical or mental torture, nor any other form of coercion, may be inflicted on prisoners of war to secure from them information of any kind whatever. Prisoners of war who refuse to answer may not be

threatened, insulted, or exposed to unpleasant or disadvantageous treatment of any kind' (*Geneva Convention (III)* ... 1949, §I, art. 17).

6. To put the techniques used in Guantánamo Bay into further historical perspective, Rejali surmises that they are ultimately rooted in 'old West European military and police punishments (Anglo-Saxon modern) or they have descended from the pre-World War II practices of French Colonialism (French modern). Most techniques are low-tech, many rooted in native American policing going back to the nineteenth century' (Rejali 2007, 378).

7. A relatively new model of psychological torture that does not mark the body of its victim, in which music is deployed to inflict pain. '"No-touch torture" using music to dissolve others' subjectivities has been imposed on persons picked up in Afghanistan, Bosnia, Egypt, Ethiopia, Gambia, Indonesia, Iraq, Mauritania, Pakistan, Thailand and the United Arab Emirates' (Cusick 2006). For Cusick, this strategy entails that 'a detainee ... must experience himself as touched without being touched, as he squats, hands shackled between his shackled ankles to an I-bolt in the floor, in a pitch-black room, unable to find any position for his body that does not cause self-inflicted pain ... the experience creates a nexus of pain, immobility, unwanted touching (without-touch); and of being forced into self-hurting by a disembodied, invisible Power' (ibid.).

8. As verified by Roman Vinokur, 'acoustic noise starts inflicting discomfort to the ears at 120 [decibels] and pain at 140 [decibels] in the audio region. Eardrum rupture occurs at approximately 160 [decibels]; lung rupture may happen at 175 [decibels]' (Vinokur 2004, 23).

9. An example of the disembodying capacity of waveforms is evident in Alexander Graham Bell's invention of the telephone—a technology that, he conceived, would allow him to speak to his brothers when they passed away. Henriques also notes that 'in his *Gramophone, Film, Typewriter*, Friedrich Kittler gives a fascinating account of how the first use for phonographic voice recording was to listen to the literally disembodied voices of the dead' (Henriques 2003, 461).

10. R. Murray Schafer has consistently denigrated the transmissions of 'noise' and excessive levels of sound by humans and has conversely sought to record those sounds perceived to be under threat of being silenced. As a result of these convictions, Schafer ran the acoustic ecology studies in the late 1960s at Simon Fraser University in Vancouver as part of his World Soundscape Project.

11. The meaning of *schizoid* differs from the meaning of *schizophrenia*. The two are related, however, in terms of affected subjects showing signs of social disconnection from their environment and from those inhabiting it. Schizoid personality disorder is a malady attributed to those subjects who live in solitary conditions, with few relationships, and display a perceived lack of empathy towards others.

12. A subtype of schizophrenia also identified as disorganised schizophrenia.

13. In psychotherapeutic terms, *reality distortion* occurs when a subject experiences hallucinations and delusional episodes that lead to a distorted perception of their existence, whereas *psychomotor poverty* occurs when a subject experiences a poverty of speech, an inability to act spontaneously and a subduing of emotional characteristics.

14. Traditional Islamic songs that generally do not contain music made by instruments but that employ percussive elements. Contemporary nasheeds, however, are allowed to contain music made from a range of musical instruments in their compositions so long as the content is structured around Islamic narratives.

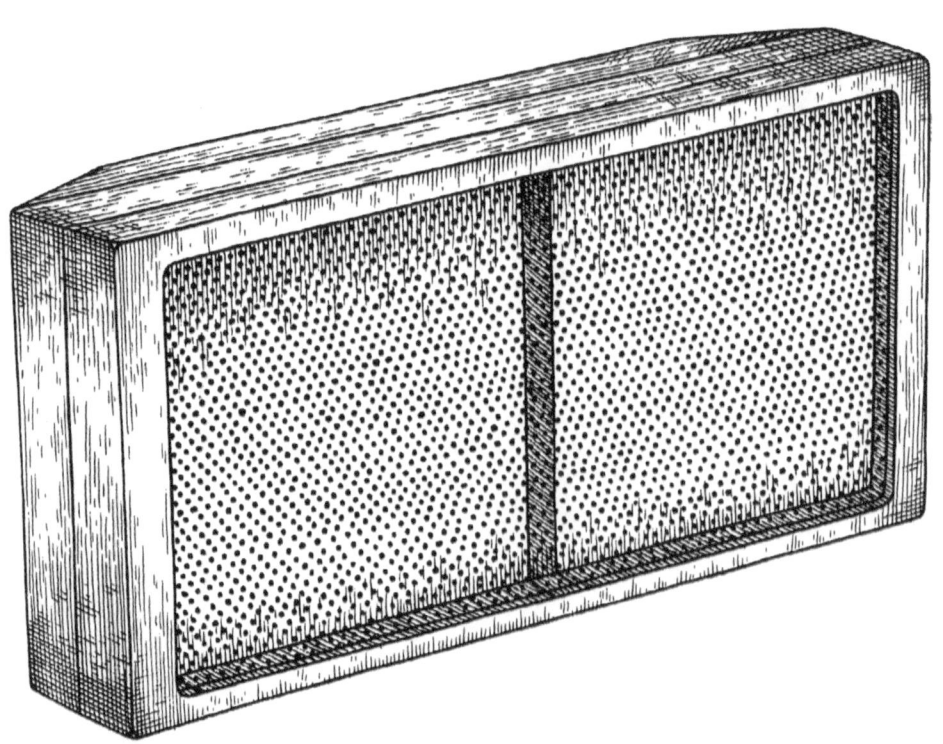

American Technology Corporation HSS-H450 HyperSonic Sound Technology Directional Loudspeaker.
Illustration by Krystian Griffiths

Chapter Four

The Covert Aims of Directional Ultrasound

HETERODYNING LOGIC

The focus of this chapter is a recently developed ultrasonic-beam technology known as either *HyperSonic Sound* or *Audio Spotlight*. The dual nomenclature results from two different companies at the turn of the twenty-first century having simultaneously developed similar technologies that aim to produce the same effect: to make sound highly directional by harnessing ultrasonic frequencies to carry it. Up until these companies recently ceased manufacturing the speakers, the LRAD Corporation in San Diego was responsible for producing the HyperSonic Sound (HSS) technology, which, in the words of their website, 'gives you the ability to direct sound exactly where you want it' (LRAD Corporation 2009). Meanwhile, Holosonic Research Labs, Inc. (referred to as Holosonics) from Massachusetts is responsible for having produced the Audio Spotlight, which, they claim, will 'add sound and ... preserve the quiet' (Holosonics 2009). Thus far the differences, if any, between the two technologies are not clear; the divergence between that of the LRAD Corporation and Holosonics only becoming apparent upon further research into the intended end users and the type of circumstances in which the systems are used.

While the LRAD Corporation geared its speaker systems towards some commercial uses, their technologies were more often utilised by military and policing organisations for the purposes of crowd control, area denial (in sites of conflict) and ship-to-ship communication. The Holosonics technology, meanwhile, has predominantly been purchased by commercial and culturally oriented organisations, such as museums and art galleries. This book's ongoing theme of the connections between cultural and military organisations is again amplified here, as the technologies employed by these institutions project ideologically related trajectories—one of the most definitive paths being the targeting of the individual body rather than the mass social body, marking an important shift in speaker-system functionality, perceptibility and location. Never before have covert frequency-based technologies such as these been employed in public, leisure and war zones to dissociate the individuated waveformed body from its environment.

HYPERSONIC SOUND

The HSS is best described by its evangelical inventor, Elwood G. Norris, in an interview he undertook in 2003 with *New York Times* journalist Marshall Sella. He is impressed enough by the invention to proclaim that it represents the first 'revolution' in acoustic technologies since the invention of the loudspeaker some seventy-eight years earlier. During the interview, Norris reverentially describes the process that allows the small flat speakers—which are connected to CD or MP3 players and an amplifier—to aim sound in highly directional beams of up to 450 feet at a consistent volume level. Sella writes:

> At the source, in the circuitry of the emitter, audio frequencies are 'stirred together' . . . with ultrasonic frequencies and then sent out as a 'composite frequency' that is inaudible to the human ear. The sound 'hitches a ride on the ultrasonic frequency', Norris says, which travels in a laserlike beam in whatever direction it is pointed. 'And here's the beauty part', he says. 'The air molecules themselves convert this ultrasonic frequency back down to a frequency that can be heard'. So unlike sound that travels on radio waves and has to be converted by your stereo's receiver, you simply need to be standing in the path of an HSS beam in order to hear the sound'. (Sella 2003)

Thus the localisation of the sound can be realised within a subject's interior physicality, as the audible element of the beam is only converted into sonic form when it touches the surface of the targeted skull, while those outside of its path hear very little. The article goes on to reveal that, from its inception, the HSS has elicited similar responses from all who experienced its directed transmissions: the lucid but insidious proclamation that 'the sound is inside my head' (ibid.).

AUDIO SPOTLIGHT

We now turn to the Holosonics website, which succinctly outlines the Audio Spotlight speaker system's ultrasonic process as 'an ultra directional audio technology invented by Joseph Pompei while studying and working in the MIT labs' that 'uses a beam of ultrasound as a virtual acoustic source, enabling unprecedented control of sound distribution'. The system 'creates a tight, narrow beam of sound that can be controlled with the same precision as light' (Holosonics 2009).

A more in-depth explanation follows, explicating the physical relations between sonic and ultrasonic waveforms: 'The ultrasound, which contains frequencies far outside our range of hearing, is completely inaudible. But as the ultrasonic beam travels through the air, the inherent properties of the air cause the ultrasound to change shape in a predictable way. This gives rise to frequency components in the audible band, which can be accurately predicted, and therefore precisely controlled. By generating the correct ultrasonic signal, we can create, within the air itself, any sound desired' (Holosonics 2019b).

With regard to the technology's capacity to function over distance, the frequently-asked-questions section of the Holosonics website informs us that the company has 'had systems mounted eight stories up on a building rooftop, beaming clearly audible sound to street-level in Manhattan' (Holosonics 2019a). In a similar vein to the 'revolutionary' claims made about the HSS, Jennifer 8. Lee, writing for the *New York Times*, contends that

> the real revolution of the acoustic beams lies not in the circuit boards but in the mind. The audio spotlight will force people to rethink their relationship with sound, as the arrivals of the phonograph, the telephone and the Walkman have done before. An occasional cathedral or dome delights us with acoustic tricks played by the architecture when sounds from far away seem to originate nearby. But those are isolated effects.
>
> With the exception of Walkmans and headsets, sound is public, a shared phenomenon. We are skeptical of those who claim to hear sounds and voices that we can't hear. Humans are immersed in a world of overlying spheres of sound. We can close our eyes but can't shut off our ears. (Lee 2001)

As Lee insightfully suggests, ultrasonic beam technology ruptures and reconstitutes the very organising principles of the waveformed domain, from the nebulous mesh of frequencies we exist within every day to a new, dissected spatiality that carves up the logic of Enlightenment perspective and our place within it. For Steve Goodman, such a schism in the formatting of waveformed territorialisation makes the Audio Spotlight and HSS technologies 'perhaps the most significant phase shift in capitalism and schizophrenia since the invention of the loudspeaker. It scrambles McLuhan's classic analysis of the opposition between acoustic and visual space, in which acoustic space is immersive and leaky, whereas visual objects of perception occupy discrete locations. Holosonic control shifts us therefore from the vibrational topology of the ocean of sound to the discontinuous, "holey" space of ultrasonic power' (Goodman 2009, 186).

The implications of this shift in episteme will be further considered throughout this chapter. As we shall see, the effect of ultrasonic linearity on the networked relations between body, environment, perception and cognition announces a new type of ambient violence. No longer can the previously trusted sonic dynamic of the echo be relied upon to locate and navigate; now the unheard inversion of this acoustic dynamic demands attention—an ultrasonic reverberation that transforms the reference points of lucidity as much as those of orientation.

ULTRASONIC UTILITY

We have become accustomed to employing the sonic to assist us in all facets of civilian, military and industrial life. HSS technology may signal a new way of thinking about the dynamics of the waveformed terrain, but the harnessing of ultrasound has been prevalent in the domains of science and medicine for decades. The physically penetrative nature of ultrasonic waves—which can pass through metal, stone and brick without causing any damage to their material structures—means that it is applicable to a number of problems that occur in industrial contexts as well. While its capacity as an inaudible and noninvasive procedure is more commonly associated with its use on the body—medical applications such as imaging and the sterilisation of surgical instruments—a wide range of more-esoteric applications and lesser-known techniques depend on the use of ultrasound. In sonochemistry, high-intensity ultrasonic waves are used to create physical and chemical changes in materials, in agriculture to measure the thickness of fat layers on pigs and cows as part of livestock management, in oceanography for tracking submarines, mapping the contours of ocean beds and searching for schools of fish. Such waveforms are also employed in systems of high-intensity cleaning, machining and flow metering and for the nondestructive testing of a huge list of industrial and commercial applications. The functionality of ultrasonic frequencies within civilian contexts is even remarked upon by Paul Virilio, who considers its applicability to surveillance techniques endemic to an array of state apparatuses aiming to render the body visible in public space. Back in 1994 Virilio reported that 'now, with ultrasound, we can bring up the image of a person who's just a tiny speck the size of a pinhead on a video tape, even if they're at the back of a dark room' (44).

INFRASONIC UTILITY

The unheard range of frequencies below twenty hertz, known as *infrasound*, has profound yet often misunderstood effects on the human subject. It is useful to remind ourselves of these effects so as to fully comprehend the wider spectrum of spatial politics within which unsound places us. Rather than being conducted by the ear, infrasonic frequencies affect us physiologically via intense silent waves of pressure that are perceived as pure vibration upon the body and within the body's cavities. If the intensity of such phenomena were perceivable by the ears, it would manifest as painfully loud sound, which, registered by the neocortex, would place the listener into an anticipatory fight-or-flight mode. Given that inaudible infrasonic waves are not transferred to the neocortex and are thus not constituted as threatening stimuli, they do not trigger this conditioned reaction. They do, however, actuate states of anxiety, as the sensorimotor and cognitive apparatuses search for perceptible sensory information but, upon finding none, become functionally compromised.

In an effort to understand such psychological and physiological effects, in the early 1960s the US National Aeronautics and Space Administration—NASA—became the first organisation to sponsor serious research concerning infrasonic influence on the body. Dr. Henning von Gierke investigated how infrasonic waves produced at rocket launch time compromised an astronaut's body's abilities to function, finding that chest-wall vibrations, alterations in respiratory rhythms and gag reactions all occur frequently upon exposure to them (his findings were later confirmed in Beaupeurt, Snyder, Brumaghim, Knapp 1969). Also discovered during these investigations (and further explored in Tandy and Lawrence 1998) was that the human eyeball resonates at around eighteen hertz (eighteen cycles per second), meaning that a 'smearing' effect transpires when such frequencies are generated, leading to possible hallucinations in the peripheral visual field. Potentially the most damaging for the body, however, and the most profound in terms of somatic effects is the seven-hertz frequency—the frequency of median alpha waves within the brain[1]. It is also the resonating frequency of the body's internal organs, meaning that upon high-intensity exposure, the body is vulnerable to pain, damage and, in severe circumstances, death.

The earliest recorded experiments in the military application of infrasound include those undertaken by Thomas Edison during World War I (as documented in Petersen 2007) when Edison invented a basic detection system for aircraft, leading to further research into submarine detection and sonar techniques. But it was only after World War II that the first significant steps were taken to harness infrasound as a weapon. Working in a Marseille laboratory to develop mobile robots that could be utilised in military and industrial contexts, Dr. Vladimir Gavreau and his assembled team of scientists 'made a strange and astounding observation which not only interrupted their work but became their major research theme'. Having experienced feelings of nausea during their work, Gavreau and his team deduced that infrasonic waveforms produced by an incorrectly installed motor-driven ventilator in a large concrete duct were to blame for the malady. 'Coupled with the rest of the concrete building, a cavernous industrial enclosure, the vibrating air column formed a bizarre infrasonic "amplifier"' (Vassilatos 1996, 31).

Instantly perceiving the potential military application of this effect, Gavreau focused on constructing infrasonic whistles and pipes of up to seventy-five feet long in order to devise a controllable infrasonic weapon that would place its targets in an 'envelope of death' (ibid.). After numerous experiments, Gavreau was forced to admit that he could not develop a weapon that would function without its concentric reverberations also affecting those who deployed it. Ironically, the scientist would end up spending much of his research time and money on trying to invent infrasonic armour, but to no avail. Although ultimately Gavreau's research did not yield a 'conflict-ready product', it remains highly significant, as it recognised the potential for the employment of inaudible waveforms as controlled pressure. In perceiving inaudible waveforms' capacity to touch, oscillate and damage both buildings and bodies, Gavreau became the progenitor of a new range of unsound techniques of asymmetric warfare—practices that would emerge in an age of 'death-lite violence', and one that sanctifies extreme turbulence by declaring its weapons to be nonlethal.

THE NONSOUND OF NONLETHAL WEAPONS

Over the past century, the fictional projection of waveformed armaments has been common in films, books and music (in literature, for example, Ayn Rand's novel *Atlas Shrugged* [1957]; numerous examples in film include Alfred Hitchcock's *Foreign Correspondent* [1940], Sherman A. Rose's *Target Earth* [1954] and Edward L. Cahn's *Invisible Invaders* [1959]). But as an active class of practical munitions, nonlethal weapons in general are a relatively recent phenomenon. In the 1960s, US police forces used rubber bullets and chemical sprays to deter and subdue rioters. During the first Gulf War a second wave of nonlethal weapons was introduced into conflict zones, including lasers, sticky foams and caustic solutions. A range of acoustic weapons that could generate pain as well as infrasonic technologies causing debilitating nausea and sickness completed this new arsenal.

Nonlethal weapons have been developed and introduced into theatres of conflict with their media representation in mind. Governments, police and military leaders understand well how public support can quickly be eroded by empathetic responses to media reportage of instances of violence and pain. One outcome of their learning to take this effect into account and to play the media game has been the emergence of 'research [into] the "bioeffects" of beamed energy . . . searching the electromagnetic and sonic spectrums for wavelengths that can affect human behavior' (Pasternak 1997). Escalating this notion, it could be said that government, police and military leaders have not so much learned how to play the game as reimagined it by changing the locations on which it focuses—from the observable battlefield to the non-visible theatre of operations, from the zone reverberating with screams and explosions to an unsound environment in which one cannot hear or locate the presence of the 'enemy'.

The reasons behind the military's movement from noise to silence, from the perceptible to the nonperceptible, are ideologically composed from the age-old scores of camouflage. It is only recently, however, that weapons systems have caught up with the martial desire to outmanoeuvre and damage an enemy without risking any overt presence of one's forces or armaments. Dr. John Alexander, former member of the US Army, Special Operations, and advocate of nonlethal technologies, champions asymmetric and abstracted types of warfare on the grounds that 'there is a misconception that war is about killing. . . . War is about imposition of will. Nonlethal weapons fit in the spectrum of this' (BBC News 2003b). As noted by Steve Wright (2000), the nonlethal techniques supported by acoustic weapons are now being deployed in both military environments and civilian contexts—such as hostage rescue, crowd control and urban combat—precisely because they do not speak the traditional language of observable pain. The empowering of agency through waveforms has reached a point where the instrumentality of a weapon is defined as much by its capacity to negate its own identity and presence as it is by its potential to propagate political beliefs or enable geographical extension.

As outlined in a memorandum written by the Arms Division of Human Rights Watch, 'There are indications that acoustic weapons are also being developed for secret "special" missions and covert operations such as counterterrorism. Acoustic weapons are also being developed with commercialization in mind, for civil law enforcement, border control, and internal prison use.' Further, 'The existing military literature indicates that acoustic weapons—across the entire frequency spectrum, from infrasound to ultrasound—have the ability to cause severe pain, loss of bodily functions, and bodily injury. Depending on the frequencies, intensities (decibel level), and modulations employed, acoustic weapons could cause permanent or temporary physical damage, including damage to internal organs, interference with the workings of the central nervous system, . . . tissue destruction, hemorrhaging, spasms, acoustic fever, . . . "significant decrement in visual acuity", incontinence, postexposure fatigue, and diffuse psychological effects' (Human Rights Watch, Arms Division, 1999).

Sonic weapons force us to rethink violence and its associated cartographies of affect, pain and temporality. Geography and the extension of the self also need to be reconsidered as the taxonomy of conflict is being reformatted by techniques and tools that disrupt the traditional Western history of perception. Such technologies are instead dedicated to orchestrating a state of nonpresence. In a future where such obscured epistemologies are augmented by the modalities of nonsound and nonlethal weapons, this begs the question, *How do we begin to perceive nonpain?*

As far back as October 1992, this future had been foretold in the form of a patent: the first practical application of covertly applied waveforms came by way of Dr. Oliver Lowery from Norcross, Georgia, and his US Patent #5,159,703 for a 'Silent Sound Spread Spectrum' (also known as the SSSS, the S-quad and the Squad). The patent abstract for this silent subliminal presentation system declares it to be 'a silent communications system in which nonaural carriers, in the very low or very high audio-frequency range or in the adjacent ultrasonic frequency spectrum are amplitude- or frequency-modulated with the desired intelligence and propagated acoustically or vibrationally, for inducement into the brain, typically through the use of loudspeakers, earphones or piezoelectric transducers. The modulated carriers may be transmitted directly in real time or may be conveniently recorded and stored on mechanical, magnetic or optical media for delayed or repeated transmission to the listener' (Lowery 1992).

More recent developments of such ultrasonic technologies include LRAD Corporation's production of the high-intensity directional acoustic (HIDA) system (an offshoot technology of the HSS discussed above) and a microwave weapon by WaveBand Corporation called MEDUSA (Mob Excess Deterrent Using Silent Audio), which—echoing the objectives of the HSS—beams sound directly into a subject's cranium. According to Lev Sadovnik of the Sierra Nevada Corporation, MEDUSA 'exploits the microwave audio effect, in which short microwave pulses rapidly heat tissue, causing a shockwave inside the skull that can be detected by the ears. A series of pulses can be transmitted to produce recognisable sounds' (Hambling 2008). Navy reports concluded that MEDUSA should be considered a success given its ability to produce an array of painful effects ranging from irritation to incapacitation. Such assertions of technological accomplishment are routinely issued by military and corporate sources. Some researchers, however, remain sceptical of any claims of the ability to efficiently induce waveformed pain. Roman Vinokur, for example, says that 'much of what is published on acoustic weapons in the media (in particular, about "acoustic bullets" and "deadly" infrasound rays) is often based on hearsay and misunderstandings, leading to criticism by professional scientists' (Vinokur 2004, 19).

Widely acknowledged an expert on the efficacy of acoustic nonlethal weapons, Jürgen Altmann, recently retired from the Technical University of Dortman, has repeatedly expressed his doubts as to the practicality of acoustic nonlethal weapons (Altmann 1998 and 1999). Employing a range of testing methodologies, Altmann systematically challenged the scientific basis upon which corporate, military and government representatives founded their claims about the functionality of such technologies. Even though Altmann's views are respected, it is plausible that his position on nonlethal weapons is borne of a personal commitment to their nonproliferation and that he has a vested interest in downplaying their efficacy in the hopes that such findings will deter the military from pursuing further developments. While his findings are distinguished by his scientific acumen, it is important to realise that the actual effectiveness of acoustic nonlethal weapons is not all that is at stake here. There has been a perceptual paradigm shift in the manufacturing of affect, pain and violence. The new procedures aim to silently territorialise new imperceptible realms, but their strategies can be located in our experiences of the everyday and the everywhere.

MUSIC AGAINST YOUTH

> Could it be that property owners now have their own sonic weapon in the battle against hooded youth who have previously attacked their pacified soundscape with their voices, ring tones, pirate radio, and underground music infrastructures? Has ultrasonic warfare graduated to the High Street?
>
> —Steve Goodman (2009, 183)

Rather than discuss at length the numerous inaudible systems of waveformed conflict now operative within our urban environments, here we concentrate on one device that can be seen as representative and as implying the future domestic trajectory of such technologies. The Mosquito MK4 ultrasonic youth deterrent is an apparatus that sounds like it must be some kind of revenge technology for the Children of the Damned (Leader 1963). Indeed, the inaudible speaker system is enthusiastically marketed as such on Moving

Sound Technology's webpage selling the Mosquito MK4 with Multi-age, which, they say, 'now has two functions. Either set the device to 17 kHz to disperse groups of troublesome teenagers *or* set it to 8 kHz to disperse people of any age from areas where loitering can be an issue such as subway terminals, car parks or any areas where people feel insecure at night due to other people loitering in the shadows etc.' (Moving Sound Technologies 2018; emphasis original). Not content with identifying and profiling the sexual, economic, racial and religious body of the domestic enemy, Western culture has recently redoubled its engagement in its conflict with a particular internal nemesis: the body of youth.

MOVEMENTS OF AN UNHEARD BODY

As ultrasonic nonlethal weapons come to spatially reorganise—whether by deceiving individuals (through the covert nature of HSS beam technology) or moving 'undesirables' on (through irritation and pain inflicted by the Mosquito)—these dynamics quietly signal a new frequency-based era of influencing, manipulating and torturing the body. Whereas we could once feel vaguely safe from intimate waveformed violence in a crowded public space, today anyone, whether static or in transit, can potentially be isolated and defined as an admissible target through the application of ultrasonics. No longer do we simply move through and situate ourselves within the rhythmic spill of the speaker system. The existence of such inaudible-transmission technologies designed to trace precisely our spatial activities and deviances urges us to renegotiate the terms of presence and the role of the waveformed body in space. With the military and entertainment industries currently responsible for implementing this new frequency-based politics, it will inevitably take time to fully comprehend its coordinates, effects and perceptual taxonomy. Yet there are signs that such knowledge is being assimilated and put to work in tactical countermanoeuvres, as exemplified by children's use of ultrasonic frequencies as smartphone ringtones (as owing to the effects of age-related presbycusis, over-twenty-fives generally do not register such high frequencies)[2]. Considering the evidence at hand, it is safe to say that the technique of training the unperceived (the HSS beam) is equally at home on the battlefield and in the civilian metropolis and that its capacity for engaging with mobility—both its own and its targets—knows no boundaries.

In *Speed and Politics*, Virilio asserts that an essential strategy of warfare has always been '*the art of movement of unseen bodies . . .* able to strike no matter where and no matter when' (Virilio 2006, 62; emphasis original). By changing perceptual terrain—from the physical to the frequency-based domain—the military-entertainment complex has effectively upgraded Virilio's analysis, uploading it into a waveformed system of thought and composing new forms of nonpresence in the process. The previously camouflaged (unseen) military body is now conceptually transposed onto the frame of the victim, with the negation of any observable pain and violence, exchanging immaterial presence for a presence that resonates with inaudible intent—the body of the unheard. Quiet but focused, this somatic will extends itself through the HSS's in-ear perspective. Aware that 'for the sonic unconscious, speed of movement is processed directly in the body' (Fuller 2005, 30), the perceptual shift projected by the HSS's unheard body is rationalised by a more clandestine instrumentality: the aim to deceive, disorient and decentre by using speeds and movements of information that are directly processed in the brain.

A WHISPERING PARASITE AND SIAMESE CONSCIOUSNESS

The HSS's modus operandi is to psychologically deceive those in its covert trajectory and to amplify its inaudible potential to decompose the perceptual rationale of its target. The technology's facility to not only speak to the inner voice but also more precisely create another internal articulation could be considered a central objective of the speaker system. By pressuring and multiplying the internalised voice of a subject, those employing the HSS for purposes of a more insidious nature seek to psychologically disorient, in turn creating an excessive sense of self. The ultrasonic beam's prospective channelling of the schizophrenic condition as well as the challenge it poses to rationality are crucial to the overall analysis of the technology. In order to understand how ultrasonic weapons present new ways of defining the presence, agency and extension of the waveformed subject, their instrumentalisation of perception needs to be further examined.

In chapter 3 it was shown how techniques harnessing audible overload for psychological affect reached their zenith in the torture cells of Guantánamo. Through the application of acoustic repetition and excess, the detained subject is pressured by sound into deteriorating rhythms of psychological collapse and breakdown in the hope that the detainee's inner voice can be located and amplified. As mooted above, the ultrasonic beam presents new ways of thinking about perception, excess and agency. Instead of perceptible sonic pressure, it involves an imperceptible channelling of externalised agency, one that challenges the efficacy of the sonic as the sole manifestation of waveformed force. In the world of ultrasonics, the power of 'excess' no longer resides in the external production of sonic dominance and its reverberatory politics. In ultrasonic terms, the operative properties of excess are now remodulated to directly manifest and propagate themselves within the internal cognitive facilities of the subject as voices are beamed directly into a target's cranium.

The extension of one's voice into the mind of another, without its being perceptible by the external sensorium, circumvents all traditional definitions of the self's relationships to the world at large. This transmitted voice is not identified as emanating from an external source, however. Rather, it is deceitfully projected in the hope that it will be perceived as an internally occurring presence. The HSS ultrasonically simulates a secondary self, a whispering parasite that engages with a target's inner voice to spawn a siamese consciousness.

CHANGING THE CHANNELS OF DUPLICITY

Inverting the traditional military roles of deception and camouflage, the HSS performs a neural charade within the new battlefield that is the skull. Virilio signalled this martial channelling of duplicitous affect into the psyche (without going so far as to name its anatomical destination) some time ago. Commenting on one of the prime directives of conflict—the will to impose one's perceived reality on another—he traces the way strategic deception has been reconstructed according to the dictates of technological progress. He writes, 'War was always linked to perceptual phenomena, such as I call the "logistics of perception". The technologies are such that it no longer suffices to camouflage a plane but instead its path must be camouflaged to conceal its movements by means of disinformation (deception) that fabricates false random trajectories. The ruses of war are as old as war, except that today the deception is in images, radar signatures, electronic countermeasures' (Virilio 2002, 33–34).

Noting the increasing significance of the covert trajectory and the subliminal channel, Virilio explicates the contemporary dynamics of applied violence and their evolution through waves of imperceptible pressure. In doing so, he articulates an emergent politics of ultrasonic instrumentality, extending the amplitude of his research, which had previously overlooked, and underlistened to, the significance of waveforms. Recounting previous ocular techniques of deception, he references the camouflaging of aircraft, but there are many other historical strategies that could be referenced instead. The recent history of English and American military practices of waveformed deception is rich and varied. The following brief outline of some of these techniques should serve to further expose the perceptual shift instigated by the ultrasonic beam, a move from practices of audible duplicity to the undetectable insidious motility of nonsound.

During the final days of World War I, German forces regularly intercepted radio transmissions, which allowed them to strategically anticipate American military manoeuvres. It was in this context that two American soldiers successfully introduced the term *voice scrambling* into the lexicon of techniques for effective communication. Realising that 'in warfare, knowledge must be complemented with deception' (DeLanda 1991, 185), Ben Carterby and Private First Class Mitchell Bobb simply spoke their native Choctaw language to each other and in so doing created a closed channel of information dispersal for which they became known as *code talkers*. More than two decades later, during World War II, the success of this strategy would be remembered and reenacted when the US military employed four hundred Native Americans (of Navajo descent) as code talkers. Since the Navajo language was articulated through its oral tradition, this meant that there were no written materials to be captured and deciphered. Adding to the unintelligible nature of their transmissions, the Navajo code talkers created new words that even other fellow Navajo would not know, resulting in a message-relaying system that proved impossible to decode. As detailed by Deanne Durrett (1998), the ideological practices of secure sonic-communication systems had begun in earnest.

Researched in detail by Philip Gerard (2002), the following example of military duplicity also occurred during World War II. The 'Ghost Army' was a US tactical deception unit consisting of approximately 1,100 artists, actors, musicians and other 'creative types' sequestered from art schools and advertising agencies to construct fake military installations, rubber tanks, fabricated soundscapes (acoustic techniques that were later used in the first Gulf War, when speaker systems placed behind sand dunes dictated the movements of the battlefield by transmitting imaginary battle sounds) and false radio transmissions. The directive of the Ghost Army was to saturate the Nazis with duplicitous acoustic signifiers of mass troop and artillery manoeuvres (mounted on armoured vehicles, large speaker systems transmitted soundtracks in covert locations so as to suggest the presence of mass numbers of soldiers and equipment) and to more generally transmit disinformation about the numbers, plans and whereabouts of the Allied forces. Conversely, during this period of time, the Nazis utilised sound in the form of coerced music to acoustically camouflage the grotesque cruelty of their 'Final Solution to the Jewish Question' in Europe. In the Birkenau concentration camp, 'the camp band had regularly been required to play adjacent to the railway platform during selections or to perform in front of the gas chambers, thereby deceiving newly arrived prisoners into believing that they did not confront an immediate threat to their lives' (Fackler 2003, 117).

The camp musicians were also forced to cover the sounds of extermination, as recalled by trumpeter Herman Sachnowitz, who recounts that 'we also played on other occasions, especially during executions, which usually occurred on Sunday afternoons or evenings . . . perhaps they intended to drown out the last protests and final curses with music' (ibid., 116).

The final example of sonic deception is more recent, one that has been used during all major armed conflicts led by the British and US armies since World War II. This is the clandestine radio programming of the military's psychological-operations division—or psyops. By fabricating phantom radio stations able to move their transmitters at will (and to broadcast over long distances), military organisations are able to channel disinformation and propaganda directly into the homes, workplaces and battlefields of the countries they are engaging within. Furthering the range and scope of transmissions, military aircraft such as the EC130—according to the *Psywarrior* website (2010)—have been 'converted to flying radio and television stations, capable of preempting a country's normal programming and replacing it with whatever informational broadcast that is felt necessary to get the message through to the listening audience'. Given the success of the duplicitous radio stations, they have become an integral part of any contemporary conflict strategy, as proven by their deployment in Grenada (1983), Panama (1988–1990), the first Gulf War (1990–1991), Somalia (1992–1993), Haiti (1994), Bosnia (1995) and Iraq (1998).

The US government's use of unconventional modes of warfare via guerrilla radio stations established its most persuasive and intoxicating rhythms of deception, however, during the Vietnam War. During this conflict, a seven-channel network carried clandestine 'black' transmissions by the United States that falsely claimed to be from Communist rebels. Inciting suspicion and fuelling hatred, the network relayed fraudulent information to the North Vietnamese, allegedly exposing the 'frightening' and 'dangerous' desires, politics and objectives of the Communist threat. In this case, it could indeed be said that 'sonic deception therefore emerged out of the power of audible vibrations to generate an affective ecology of fear' (Goodman 2009, 42). Considering that clandestine radio technology has consistently featured as a psyops strategy in all US–led invasions since World War II, it is safe to say that, within the taxonomy of sonic deception, it rates as one of the military's most potent techniques. Its efficacy is reliant on its capacity to directly transmit disinformation to citizens and military personnel alike and in its potential to instigate mass transformations in a culture's sympathies and territorial behaviours. Also not to be underestimated is its ability to broadcast waves of dispiriting anxiety, an important asset if it is to be believed that 'the *field of battle* is a *field of perception* which must be organized in such a way as to control the movements of the adversary and cause them to follow a false lead, to demoralize them and exterminate them' (Virilio 2002, 96; emphasis original).

THE PSYCHOGEOGRAPHY OF A SCHIZOPHRENIC'S EAR

Transmitting channels of disinformation from hidden locations, clandestine psyops radio shares certain ideological trajectories with ultrasonic-beam technology. It is in their scope that they differ. Whereas the radio focuses upon creating multiple mass deceptions of the social body, the HSS targets the individuated body and endeavours to alienate and isolate it from its enveloping social networks. The convergences of these technologies can easily be registered again, however, in their dislocative schema: The capacity of both radio and the HSS to induce multichannelled states in or around their targeted listeners is reliant upon the inability of the auditor to locate the source of the transmission—a schizophonic or acousmatic state.

R. Murray Schafer's notion of *schizophonia*, 'the split between an original sound and its electroacoustical transmission or reproduction' (Schafer 1977, 90), has been mentioned in previous chapters. It is only in the ultrasonic-beam technology's rupturing of audible agency, however, that the full significance of this term for our subject can be fully recorded and analysed. First coined in 1955 by Jérôme Peignot and Pierre Schaeffer as a term for how one listens to musique concrète, schizophonia would later be conceptually remodelled by musicians, sound artists and sonic theorists. In 1994, film theorist and composer Michel Chion, reconceptualising Schafer's notion of schizophonia, called audible sound without an observable source *acousmatic*. Noting the reflections on the sonic domain that led Chion to coin the term, 'Chion', Paul Moore writes, 'points out that sound in everyday life is omnidirectional, composed of both direct and indirect reflections and many pieces of aural information' (Moore 2003, 275).

In his writing about the historical instigation of this schizophonic state of sound, Schafer referred to the 1870s, the decade during which Alexander Graham Bell invented the telephone (for which he received his first patent in 1876) and Thomas Edison the phonograph (created in 1877)—the technologies that signify the beginnings of the Western fascination and drive to disembody the voice from its anatomical mechanisms. From a more general perspective, this cultural process dislocated the primacy of the sonic by relocating the perception of its original production. In the composition of such a nomadic waveformed modality, 'the separation of sound from its original source through electroacoustical technology instantly impacted the cultures of the world' (Bishop 2002, 1). Jack Bishop concludes that 'this *schizophonic split* has arguably been the single most important moment in the history of music' (ibid.; emphasis original).

As much as the above quotation may be historically instructive, a new impasse has materialised in the twenty-first century. A new split occupies the waveformed domain, induced by the technological pressure applied by the military and entertainment industries—one that is perhaps of equal importance to the schizophonic rupture. This fissure is different, however, as it orchestrates several conceptual scores. The first is that of the separation of the sonic from the audible, as the sound wave is silenced and redefined in the ultrasonic weave of the HSS beam. The second is the potential directive of the beam and its quiet calibrations, divorcing the subject from its rational perception of the self and its corresponding relationships with its environment. Between the estrangement of the rational mind and the remodulation of the frequency-based terrain, ultrasonic weaponry simultaneously operates both on the body and on its extension, fabricating a new psychic space in the process.

As such, ultrasonic-beam technology either represents the final stages of schizophonia or, more persuasively, announces the evolution of a new state of waveformed consciousness organisation and agency that has yet to be named. If this is so, then one of the first observations of this incipient era is that 'hearing voices' can no longer be conceived of as the sole preserve of the religious, the chosen or the insane. Possibly anticipating that these voices would be rechannelled back into Western subjectivity, Schafer began compiling a list of those who were culturally assigned to receive such voices and explained their connection. He wrote, 'The ear of the dreamer, the ear of the shaman, the ear of the prophet and the ear of the schizophrenic have this in common: messages are heard, but no matter how clear or compelling they may be, there is no evidence of a verifiable external source. The transmission seems intracranial, from an interior sound source to an ear within the brain' (Schafer 2003, 33–34).

We can now expand this list, of course, for the HSS prospectively envelops all into a postschizophonic logic. There is no picking and choosing of receivers as a function of their religious beliefs, spiritual expectations or symptoms of psychic disorder; there is simply an inaudible directive to channel an unsound murmur into the cranium of a targeted body.

By silently calibrating the index of rationality, the HSS has begun to define a new epoch by stimulating theories about the production and predictability of waveformed (in)sanity. Beyond the application of chaos theory to stock markets in order to preempt currency fluctuations—or 'the mathematics describing the onset of turbulent behavior in flowing liquids . . . applied to understanding the onset of armed conflicts between nations' (DeLanda 1991, 57)—this is a new speculative analysis of the creation of disorder, an unsound theory that aims to delineate the sublime equations of inner turmoil rather than to comprehend the logistical disarray of territorial disputes or virtual economic crashes. It is a narrative that employs the mathematics of waveformed turbulence to articulate conflict between voices, and its predictions of future ultrasonic organisation within the military, entertainment and civilian economies of covert psychological violence are proving to be chillingly accurate.

A SILENT KILLING IN THE TRANSAUDITIVE ERA

Over the course of this book, the antenna-body has broadcast its journey from increasingly compressed architectural spaces—from the factory/public workspace to the domestic/private space of the compound to the incarcerated/ cellular space of the torture cell. Directional ultrasound carves out yet another space from which it will transmit and receive its signals: neural space. After the cellular space of Guantánamo Bay had been sonically mapped and overloaded, the next logical step for the military was to decompose the existing architectonic symphonies of violence and to move the theatre of operations directly into the skull. With no operating costs, no visible aberrations and no escape velocities to worry about, it was a logistically sound and predictable transfer. The emergent architecture of this migration will not be constructed from concrete, metals and plastics; instead, it is cranial, engineered from neurons, synapses and transmitters. Its driveways and highways are no longer laden with asphalt and traversed by motor vehicles; its routes are axons, negotiated by signal pulses called *action potentials*. The focus of ultrasonic territorialisation has opened up a space that, in bypassing existing material constraints, opens up a new frontier in frequency-based conflict. As Goodman predicts, 'The colonization of the inaudible, the investment in unsound research, indicates the expanding front line of twenty-first-century sonic warfare' (Goodman 2009, 187).

Acknowledging the development of these new spatial dynamics, the HSS's 'voice-to-skull transmission' negates the need to define an acoustic wilderness or to construct walls of sound and instead employs unsound methods to engage directly in a construction, deconstruction and reconstruction of inner space. As suggested, it is via ultrasonic technology that the military-entertainment complex shifts its operations and apparatus from more-predictably structured audible locations to new acoustically nebulous zones that facilitate asymmetric empowerment and arrhythmic agency. This break from operating solely in the sonic environment is indeed logical for a military that has also become vulnerable to the intense connectivity of networked media culture. As the civilian populace acquires access to increasingly effective communications technologies (and systems of knowledge mobilisation), its members also become more able to independently run resistant interference patterns across the waveformed and visualised global infoscapes via the Internet, cable technologies, satellite systems and pirate radio. Accordingly, it has become essential for the military-entertainment complex to resituate selected dynamics of engagement into perceptually inchoate and intangible topologies.

In chapter 3 it was argued that both the presence of the captive within Guantánamo Bay and the presence of the camp itself were manipulated by an array of perceptual techniques employed to simultaneously amplify/expose and mute/secrete the detainee to and from the rest of the world. This situated Guantánamo Bay somewhere between the observed and the unobserved, between the recorded and the unrecorded, in accordance with its sociopolitical/physical status as an illicit offshore torture haven. Guantánamo was an external environment where new research could be undertaken and transmitted in the name of the state—a governmental 'escape architecture' that evaded both moral and legal responsibility owing to its nonpresence on home soil. The operating logic of the peripheral is what is important to comprehend here: this logic functions in its fullest capacity when it is manifested in the hinterlands of perception, legality and humanity. Such an ideology of the periphery is useful to the military because it empowers it to slip in and out of space, presence and character at will. The HSS extends this thinking, but its technological application induces a schism, as the literal nature of airspace becomes the new conflict zone and frequencies the ammunition through which it is negotiated, territorialised and dominated. If Guantánamo Bay represents a physical manifestation of the perceived and the unperceived, the heard and the unheard, then the HSS represents an otherworldly variation on the same theme. Between stations, the ultrasonic beam reorganises the waveformed domain via the transmutation of vibrations and codifies our perceptions by manipulating the principles of physics.

With the functionality of architectonic noise having been perfected within the Guantánamo cells, the will to perfect the violent utility of silence has been mandated as the next step for those interested in commanding zones of conflict and leisure. By achieving this shift, the military-entertainment complex will become adept at seamlessly transposing their strategies and techniques between silence and noise. Operating in the smooth space of their chosen dislocation, they are able not only to modulate their presence between the civilian zone and the battle zone, but more than this, they can oscillate between states of conflict and entertainment and ultimately manoeuvre in between the senses. Thus, while Guantánamo represents the denouement of the story of one waveformed technique, it also symbolises an illicit system of transfer. It has been suggested that the military-entertainment complex's strategic shift from noise (symbolised by the sonic torture in Guantánamo Bay) to silence (signified by ultrasonic beam technology) has been defined by its unidirectional flow from one state to another; it is the regulation of this flow and the attendant capacity to reverse at will one's allegiance to a particular waveformed state that truly represents the mastering of spatiality, perception and presence.

By moving into the imperceptible spaces outlined by the HSS, the military has mapped out a new environment within which it can exist while logistically concealing its functionality, agency and systems of extension. Consequently, it can be powerfully effective without its manoeuvres ever being perceived as such. Military organisations have always been cognisant of the power residing in those phenomena that are not perceivable by the sensorium. Now that technology allows them to inaudibly articulate the finer points of this supposition, it is necessary to rethink Lex Wouterloot's contention that 'it is not to be expected that in the near future the battlefield will become silent. The transauditive era of the silent killing has not yet dawned' (Wouterloot 1992). Decades after this statement was penned, we find ourselves beyond the cusp of this era, enveloped within the dual vibrational dynamics and ideologies of an unheard temporality. In this epoch, silent killing is as much a prime directive as noisy death. When it is useful for the act of killing to be heard, the military's actions will be audibly resonant. When it is necessary for this process to be silent, the weapons employed will be those that can simultaneously mute the target as well as the environment in which it previously existed.

As has been deduced, the waveformed manoeuvres of the military-entertainment complex are not totalising ones. The military-entertainment complex leaves many of its techniques, apparatus and signifiers of violence within the observable realm so that, should its power, influence and threat require observation in order to be deemed effective, then perceptible channels are left open through which it can transmit its agency. Just as 'the power of the countermeasure . . . resides in its apparent nonexistence' (Virilio 1994, 66), part of the power of an attack resides in its inhabiting the realm of the perceptible. To extend this logic, whereas presence is represented by noise and nonpresence is symbolised by silence, together they signify a thirded vacillating demeanour that discredits and updates Jacques Attali's declaration that 'today, Noise triumphs and reigns supreme over the sensibility of men' (Attali 1985, 23). From a military and entertainment perspective, neither silence nor noise is triumphant over human sensibility; rather, what dominates current strategic thinking is the potential to inhabit either state while harnessing its waveformed dynamics to apply force and pressure.

WE ARE NEVER OUTSIDE OF GEOGRAPHY

The manner in which perceptions of the frequency-based terrain are disoriented by ultrasonic linearity reveals a crisis of spatial reasoning that breaks down old divisions within the waveformed environment. In this new context, domination is not accomplished by entering another's space and plundering their land but, rather, is expressed by the ability to pass through such space in order to invade another's skull. Through covert positioning and deceptive synthesis, the HSS disrupts apparent relationships between the concatenation of events linking the point of transmission with the point of audition and in doing so challenges the traditional assumption that 'the locations of sounds you hear are connected with the locations of their sources in the environment' (O'Callaghan 2009a, 5). As this ultrasonic logic of incoherency compels the composition of new ways of thinking about waveformed location and situation, it concurrently gives new voice to the language of an age-old struggle. The conflict over territory (whatever its formulation) has been resonant within human- and animal-kind ever since environments could be traversed, as pointed out by Edward Said in *Culture and Imperialism* (1993) wherein he claims that we are never outside of this negotiation of geography and, by extension, spatiality.

Virilio tells us that 'social privilege is based on the choice of viewpoint (before attaching itself to accidents of fortune or birth), on the relative position that one manages to occupy, then organize, in a space dominating the trajectories of movement, keys to communication, river, sea, road, or bridge' (Virilio 1977, 94). From the perspective of those deploying the ultrasonic beam, the ocular viewpoint is still significant, as targets need to be sighted in the first instance in order for the HSS to be trained upon them. But it is the waveformed channel that privileges the user with the key to discreet communication. The linear transmission of the frequencies transgresses the mapped logic of distance and vehicular transportation by generating concealed passages that aim to violate the rationale of the target's interior spatiality. In this way the ultrasonic beam renders the connection between space and distance a tenuous one.

Within this emergent ultrasonic environment, the subject finds itself brokering relationships with other bodies, both individual and en masse, in predictably divergent ways. The dynamics of these interactions are suggestive of Georg Simmel's (1903) notion of formatting and maintaining distance as a metropolitan strategy. Thus 'social distancing, as a type of performance in response to the overbearing rhythms of the city, was thought by Simmel to characterize the urban personality' (Allen 2000, 61). It is just this kind of fear—of being engulfed in swells of urban orchestration—that echoes the solipsistic aim of the ultrasonic beam. It could be said that the HSS has embraced the romanticised continuum of modern urban alienation and recomposed its inherent distances into a hypermodern streamlining of isolation. The desire of urban subjectivity—to be listened to and recognised as a differentiated individual—has been taken to heart by the instrumentalisers of an inaudible weapon that ultrasonically amplifies this will and extends it by disembodying the self, detaching it from its social network.

Supporting and contemporising Simmel's analysis of the disconnected urban dweller and her remote associations with the city and those who inhabit it, John Allen ascertains that 'to bring Simmel up-to-date in this respect, perhaps there is now as much a need to live the "global" intensity of relationships and their effects at a distance as there is the complex rhythms within cities' (Allen 2000, 63). Such a statement harkens back to the 1980s and to the harmonic resonances of the British and American conservative governments of the time and their individualistic ethos, which pervaded all strata of public and private life, the most pertinent evidence of these sentiments being Margaret Thatcher's relentless efforts to break up collective-labour associations such as the trade unions (as recorded in Dorey 1995). In today's era of globally networked consciousness, social isolation has become relatively difficult to achieve. In terms of labour, the remote affordances implied by the network are expressed in the decentralisation of the urban subject, who can now work, live and learn at distance from the city.

Teaching, receiving information and interacting remotely are for robotics theorist Ken Goldberg (2000) areas of concern for telepistemology—an upgraded electronically networked version of epistemology that aims to reveal current practices of accruing knowledge through mediation. If, in making such removed behaviours possible, the Internet symbolises the potential of infinite connection, in contrast the HSS technology represents the dystopian reverberation of Western science, dealing as it does in severance, detachment and mutation. Whereas the Internet transmits the bifurcations and minutiae of informational vicissitudes over distances and from a distance, the HSS pitch-shifts telepistemology's mandate into a paranoid echo of videodromotic transmission[3]. For it does not take a great leap of association to suggest that the arcane dark signal in the 1983 film *Videodrome*, causing neural transfiguration and hallucination, has technologically evolved from a cultivated science-fictive blip in 1983 into a martially distributed channel of mental destabilisation and spatial dislocation in 2019.

In the HSS's sounded-out terrain of obfuscation and echo, the transformation of a cultural pulse into a martial channel suggests that cultural abstraction and military camouflage were always meant to meet within such ultrasonic topologies. If abstraction is concerned with distance (removing oneself from the language of the immediately definable) and camouflage is concerned with deception (the blurring of the immediately definable), then the HSS's hushed sibilance is intimately obliged to dislocation and displacement (the fracturing of the immediately definable). For Virilio, 'the notion of displacement without destination in space and time . . . imposes the primordial idea of disappearance in distance and no longer in the danger of cataclysm'. In describing the ambiguous presence of the contemporary war machine, it posits it in a state of terminal transfer as 'it rushes nonstop toward the beyond' (Virilio 1977, 64). The efficacy of the HSS lies in its negotiation of silence and disappearance in both cultural and military environments. And as for the beyond, it is where ultrasonic transmissions mutate structure and where 'deterritorialisation . . . the question for the end of this century' (Virilio and Lotringer 1997, 142) inaudibly begins.

INAUDIBLE INFECTION

Extolling the virtues of the displaced, the deterritorialised and the mutated, Michel Serres offers alternative methods for analysing the waveformed spatiality composed by ultrasonic technologies. For him, the channel opened up between the HSS speaker and the targeted receiver of the beam is of more significance than is the end result or the initial transmission. More important—in view of his theorisations about the parasite—is the mutation that the information passing through this system undergoes. Appropriately, it is the space in between sender and receiver and, more specifically, its mediating attributes that designates it as an essential testing ground for communication. Qualifying the conduct of middle space and its capacity to function as an arbiter of connection and intervention, Serres further elucidates: 'Given: two stations and a channel. They exchange messages. If the relation succeeds, if it is perfect, optimum, and immediate, it disappears as a relation. If it is there, if it exists, that means that it failed. It is only mediation' (Serres 1982, 79).

Through defining *middle* or *between* spatiality and the mutative nature of the transmission, Serres develops Marshall McLuhan's (1962) discourse concerning the nature of the medium and its inevitable transformation of the content that passes through it, concluding that noise is an inevitable presence within all acts of communication. This noise signifies the ways in which the medium (the 'channel' in the quotation above) transforms the original intent of the sender. The definition of this interference is adherent to and utterly dependent on the operating dynamics of the channel, leading Serres to surmise that noise is also representative of the parasite: Stephen Crocker, in his 2007 article 'Noises and Exceptions: Pure Mediality in Serres and Agamben' explains that 'in French, parasite can mean the unwanted noise of communication, an uninvited guest, or a life form that lives off another. It is not just any particular organism or noise but rather the appearance of the medium, which compels any given system of order to either adjust to its presence or expel it'. Impelling us to further question the waveformed dynamics of the parasite, the channel generated by the HSS constitutes not so much an inversion of the medium as a camouflaging of it, troubling the identity of the parasite and, by extension, the viral, as it functions and transforms through the medium of unsound.

This is not the silence mooted by Serres when explicating the rationale of noise. It is a new, inaudible formulation of waveformed transfer and an unfamiliar way of perceiving the virus. In the evolving channel of ultrasonic communication proposed by the HSS, noise is wrapped within silence, and its ability to transform and modify is contingent upon this symbiotic relation between the perceptible and the imperceptible. If it is true that 'music makes mutations audible' (Attali 1985, 4), then the same can now be said for silence. This represents a profound shift in our thinking of interference and the mutative function of the transfer from one entity to another. Noise is not heuristically removed or operated on, and neither is it accommodated. Instead, it is disciplined and guided by the directives of the inaudible. Thus, as the dynamics of Serres' channel persist within HSS systems (the linearity of the beam being vulnerable to obstruction and thus to external noise), the conduit is predominantly orchestrated by the subversive properties of unsound. The modifying nature of the viral subsequently finds its expression in silence as much as it does in the sonic. The idea that sound possesses the power to disturb silence is no longer a static proposition, for now the inaudible can ineluctably mutate noise.

If this is indeed the case, then one could question the concepts Serres developed, drawing on the work of Claude Elwood Shannon, 'father of information theory'[4]. Summarising Shannon, Crocker relates how 'the snow on the television set, the hiss on a tape, or a missed registration in a printing operation are all instances of noise, or parasitism. In each of these cases, the presence of the medium is registered in what would, seemingly, otherwise be a clear transmission' (Crocker 2007).

Unwilling to tow the sonic line, the HSS proposes to conceal the presence of the medium by silencing it, just as digital musical-playback technologies propose to silence the scratches of vinyl and the buzz and hum of tapes. By redirecting the efficacy of the parasite, it disavows its 'relation not to things but to relations' (ibid.) and instead demands that noise (and thus the virus) is removed from the transmission so that it can be placed directly into the consciousness of the channel's target. The parasite is thereby displaced from its old set of relations between the sender and receiver so that the modifying affectivity and excessive productivity of the virus can be set to work in a new neural network of transmutation. Setting the virus free from the messenger's single channel enables a rearticulation of the nature of the transmission and a relocation of those viral ideologies that themselves have been assimilated into significant models of distribution within late capitalism.

Further to the above, a number of cogent viral principles enter into analogous resonance between the HSS speaker system and systems of viral marketing. To begin with, they share a concentration upon the target to be affected and infected. Instead of aiming at the mass social body, both systems take precise aim at the individual. Before the era of social-media influencers such as YouTubers, viral-marketing strategies were predicated on the targeting of selected individuals on the basis of their influential status among a wider demographic. When this individual was given (infected with) goods—a pair of trainers, for example—the individual's silence was also bought. For, were it known that the owner of the trainers had been paid to wear them, then the illusion of authentic credibility would, of course, be lost. It is this individual's capacity to make telling choices and their influence on the choices of others that is being purchased. By keeping silent and deceiving the wider social group about the nature of the choice, the duplicitous personal statement would be transmitted successfully to the surrounding hosts, and they would be considered to have been successfully assimilated[5]. The camouflaging of the medium and the silencing of the channel through which the effects will travel are deceptive imperatives for both viral marketer and HSS operator if their viral transmissions are to successfully infect their targets. It is upon fabrication of these states that the true capacity of the virus is fully exercised, modulating the receiver's sense of reality by extending the sender's will into their personal network of relations and connections.

ENGINEERING THE TWENTY-FIRST-CENTURY EAR

Attempting to extricate meaning from the complex network of strategic deceptions, illicit fabrications and camouflaged infections that have been crosswired into the HSS's system, the conclusion of this chapter considers the future of ultrasonic technologies. As far as the ideologies and applications of covertly directed waveforms are concerned, it seems in many ways that we face an immediate future of uncertainty in terms of the waveformed domain, its habitual mutation and the efficacy of our sensorium within its realms. It is suggested, therefore, that the engineering of the twenty-first-century ear and its claims to veracity are undergoing a series of tests, recordings and operations similar to those conducted on the eye during the nineteenth century, as analysed by Jonathan Crary in his seminal 1992 text, *Techniques of the Observer*. Commenting on the manufacturing of technological modalities that reimagined the roles of both perception and reason, Crary writes, 'The issue was not just how does one know what is real but that new forms of the real were being fabricated, and a new truth about the capacities of a human subject was being articulated in these terms' (Crary 1992, 92).

As ultrasonic technologies such as the HSS aim to usurp notions of the real within frequency-based environments, the role of the human subject within these environments will inevitably mutate. The antenna-body will be fleshed out just as the scopic body was fleshed out in the nineteenth century after it had been born centuries before during the European Renaissance. It will be the only somatic modality able to receive and transfer the newly occurring waveformed perspectives and everyday practicalities of unsound interaction. In an anticipatory mood, Goodman (2009, 128) forecasts that the coalescing rhythms of viral ideology, military application and cultural synthesis will result in a waveformed future city that is preemptive and responsive and that will constantly modify itself and the behaviours of those who inhabit it. This does not render the ultrasonic body purely passive, though; for while it is being modulated, it is also busy modifying its sensorial mechanisms so that it might survive and proliferate in the developing fourth world of the imperceptible. If we are to accept 'Walter Benjamin's claim that in the nineteenth century "technology has subjected the human sensorium to a complex kind of training"' (Crary 1992, 112), it can only be hoped that such diligent exercise will stand us in good stead for a future that will require engagement in sensory conflict on a daily basis.

Quietly orchestrated into the cadence of the city and the battlefield, the prospective approach of directional ultrasound manifests future-facing realities. Whether in the street, on the frontline or in the cinema, everyone is implicated in its ambitions. Sensory mechanisms will need to readjust in order to perceive the ways of the silent, the asymmetric and the asynchronous. This is proposed not to overdramatise the argument but rather to address a situation that has been stealthily evolving over the past two decades. As far back as 2005, the Sony Corporation had already patented an ultrasonic device to be employed within leisure complexes, a technology that, as Ian Sample reports, could 'evoke smells, flavours and even a sense of touch in audience's brains, in the hope of enhancing the movie-watching experience. . . . According to the documents, pulses of ultrasound would be fired at the audience's heads to alter the normal neural activity in key parts of the brain. . . . "This particular patent was a prophetic invention", according to a Sony spokeswoman. It was based on an inspiration that this may someday be the direction that technology will take us' (Sample 2005).

Having predicted current fixations with ultrasonic utility, J. G. Ballard's fictive writings about the dystopian trajectories of Western technologies, and the resulting neurotic pathologies that come to infect those exposed to them, are still pertinent and revealing. In 'The Sound-Sweep', the ultrasonic future has arrived, and audible music has been outlawed and rendered illicit, so that 'in the age of noise the tranquillizing balms of silence began to be rediscovered' (Ballard 1960, 49). Preempting the shift of military-entertainment-waveformed operations from the noise of Guantánamo Bay to the nonsound of the HSS, Ballard's black-humoured attack on the stifling of audible expression is prescient. Satirising the decorum of a culture that equates being quiet with being civilised, he announces future channels of inaudible transmission such as 'radio programmes consisting of nothing but silence interrupted at half-hour intervals by commercial breaks. . . . Gradually the public discovered that the silence was golden, that after leaving the radio switched to an ultrasonic channel for an hour or so a pleasant atmosphere of rhythm and melody seemed to generate itself spontaneously around them' (48).

It is clearly understood that in Ballard's world of muted protagonists the sonic is an obsolete form of pleasure. Sound has too many associations with disorder, chaos and noise—the channels through which demons are excised. Conversely, in the realm of the inaudible, divine power reveals itself, a hushed soundscape where the God of the early Christian cathedrals communicated its presence through the embrace of infrasound and where, now, the futuristic masters of technology communicate ultrasonically, in subliminal vectors of absence. Such prophecies of tranquil(ising) otherworlds are quiet music to Schafer's ears, given his deliberation that ultimately 'the final power then is—silence, just as the power of the gods is in their invisibility' (Schafer 1993, 202). Extending the logic of Ballard's story and Schafer's theory to their conclusions, these very different perspectives converge in a future world where musical turbulence, noisy exchange and sonic interaction have been eradicated from the sensorial agenda of the human: sonic fiction at its most dystopian.

NOTES

1. Measurable by electroencephalography—or EEG—*alpha waves* produced by the brain are electromagnetic oscillations that occur between seven or eight hertz and twelve hertz. Historically they have been associated with the body in a state of relaxation (just before sleep being the most common state), but more recent thought posits them as occurring when the brain is negotiating network coordination and communication. Frequencies below seven or eight hertz are referred to as *theta waves* and are associated with creative modes of thought. Brain-wave frequencies above twelve hertz, called *beta waves*, are generated during the majority of our waking lives. They are linked to the analytical modalities of thought manifested during periods of problem-solving. 'A question of importance', says David Walonick in his

1990 article 'Effects of 6–10 Hz ELF on Brain Waves', is '"If we can electronically shift the brain-wave frequencies to alpha or theta, will a person's moods or thought patterns change to those commonly associated with those frequencies?" In other words, if we can electronically move a person's brain waves to the alpha frequencies, will they become more relaxed? Will their state of consciousness change to coincide with their brain waves . . . ?' (Walonick 1990, 32).

2. The shape of young children's inner ears enables them to hear higher frequencies that those older than they are cannot perceive. As humans grow older, their ability to hear high frequencies decreases with age. This state of deterioration starts at about eighteen years old and is known as *presbycusis*.

3. The protagonist of *Videodrome* (Cronenberg 1983) is the CEO of a cable station who discovers that a signal broadcasting extreme acts of torture and violence is causing hallucinations and mutations in those watching it. A videodromotic transmission is a signal that harbours the potential to cause hallucination, violence and neural/somatic mutation in those who witness it.

4. Born in America in 1916, Claude Shannon was an electronic engineer and mathematician who conducted some of the first experiments into artificial intelligence and developed the field of information theory. By the time of his death in 2001, he was known as the 'father of information theory'.

5. In viral-marketing terminology, *assimilation* is an essential concept. One online glossary of viral-marketing terms explains that 'once a person has been exposed to an idea virus, if they internalize and retain the idea, we say that they have assimilated the meme. Typically this process comes with an amount of [re-creation] by which the individual attempts to fit the idea into their existing mental frameworks' (Zarrella 2008).

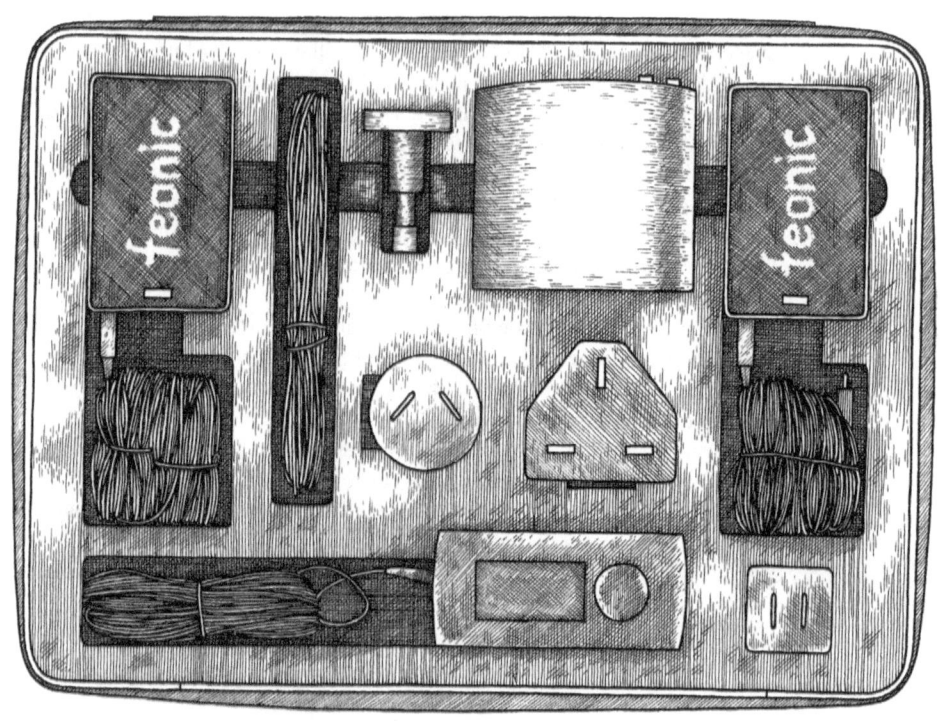

Feonic Presenter Pro–Portable PA System.
Illustration by Krystian Griffiths

Chapter Five

Whispering to Talking Windows

FEONIC: RESEQUENCING MATERIALITY

This chapter focuses upon a recently developed vibrational technology that, adorned with the somewhat insidious fairy-tale-toxic moniker Whispering Window or Talking Window, can make any flat surface, whether glass, wood, Perspex or stone, into a speaker. Two applications of the technology have been publicly tested and utilised so far, one in the context of the shop window and the second on train windows. The chapter considers both, with the shop window as the first under scrutiny. Developed back in the early noughties by Feonic Technology Ltd., a spin-off company from the University of Hull in the United Kingdom, Whispering Window is also known more prosaically as a flat-panel speaker, which denotes the manner in which the device extends the capacity and meaning of a speaker across the entire material plane. It is more technically known as a *resonance speaker audio product*. It functions by attaching a small device called a *surface transducer* onto the back of a solid surface, such as a window, wall or table, and resonating it. Music, clear-voice communication or audio effects are produced from the surface because the particles inside the surface are made to vibrate by the transducer, the energy fed into the material getting passed from particle to particle so long as there is no air gap between them.

For sound theorist Brandon LaBelle, the Feonic speaker is 'Surface Sound, a technology which enables any surface to be turned into a sounding zone' (LaBelle 2010, 188). The company's website concurs, referring to the sonic apparatus as a diffusion system that is capable of turning any surface into a flat-panel speaker and subsequently offering the intriguing premise that 'the best flat speaker has to be no speaker at all'. This renegotiation of what constitutes a speaker—traditionally a piston-based device that drives a diaphragm back and forth to emit waveforms—does have precedent. The distributed mode loudspeaker developed by NXT resonates surfaces by means of an electroacoustic exciter; it would be the obvious reference here, but presently we will focus on the Feonic device, superior in terms of both vibrational efficacy and commercial take-up. As with the HSS technology discussed in chapter 4, the Feonic flat-panel speaker proposes to change the body's relation to space, connectivity and communication in a way that requires us to think about the agency of the sound system in a completely new manner.

The component in the Whispering Window that causes surfaces to resonate—the transducer (which essentially converts energy from one form to another, from magnetic energy to mechanical energy or vice versa, for example)—contains an alloy called Terfenol-D. It is a grown crystal compound, a rare magnetostrictive material, meaning that it expands and contracts, changing in shape and size, when exposed to a magnetic field. Terfenol-D takes its name from the metals it contains—terbium (Te), iron (Fe) and dysprosium (Dy)—and from the laboratories that initially generated it in the 1970s—the US Naval Ordnance Laboratory (NOL, now the Naval Surface Warfare Center or NSWC).

Through a more historical prism, the alloy can be apprehended through its relation to the nickel and piezoelectric ceramic sonar, which was reaching the limits of its functionality in the mid-1970s. The time was right for researching and devising new materials that offered a lower-frequency range, broader bandwidth, increased reliability and enhanced acoustic power (AZoM 2002). It would take approximately another decade before a suitable successor would be manufactured efficiently via a program financed by the US Navy at the Department of Energy's Ames Laboratory; its primary application at the time was in naval sonar systems. It is only over the past decade that the general public has begun to understand the potential of this material as it is exploited in commercial designs such as the Whispering Window.

More than any other company, Feonic has been upping the ante with regard to Terfenol-D-based transducers. The first commercial application of the technology occurred with the release of their smaller SoundBug device. When the system is attached to a flat surface, it transfers an electrical signal into mechanical energy that causes the given surface to vibrate and consequently broadcast the sonic content delivered to it. This new approach to sound reproduction creates environmental sound systems from everyday objects. Early newspaper and magazine articles concerning the technology reported on how consumers reoriented its location by placing the SoundBug onto the ultimate everyday object, the body, and more specifically onto the back of the skull, with BBC's Ivan Noble reporting that 'the effect is quite strange, as the sound feels as if it is coming from inside the head, behind the ears' (Noble 2002).

Similarly to the American Technologies Corporation's claims about the state-of-the-art nature of their HSS directional speaker, Feonic's marketing asserts that the company's innovations will 'change the way you think about sound'. As an ambiently disruptive proposition, the speakers perform a waveformed machination similar to that of the HSS by simultaneously replicating the situated efficacy of the inner voice along with the disorienting ripple of the aural hallucination. The manner in which it achieves this affect, however, is another matter altogether. As opposed to the targeting and isolation applied by the HSS, with Feonic's technology, distance is negated, and the body itself becomes a potential sounding board, along with everything around it. In this resequencing of materiality, oscillation becomes the distributed agency that binds bone to graphene, skin to silica, through a choreography of waveformed standardisation.

In 2003, high-end department store Peter Jones in London's Sloane Square became one of the earliest purchasers of the Whispering Window technology. The visual merchandising manager at Peter Jones reported an increase of 49 per cent in passersby stopping to look at the store's window display, inquisitive as to the source of the music and spoken-word content emanating from it (Associated Press 2003). Given the flat-panel speaker's power to arrest the attention of pedestrians, Peter Jones went on to employ the system for its store windows in other cities, including Edinburgh, Glasgow and Nottingham. Equipped with a capacity to monitor the level of street noise, the furtive sound system is subsequently able to calculate the volume that it needs to transmit at in order for its emissions to sit just above the ambient sonic environment that it sits in and interfaces with. Sonically dressed, the window becomes a subtle attractor, a technology that pitches the surrounding atmosphere somewhere between the fiscal intimacy of the 2002 Spielberg film *Minority Report*'s personalised advertising and the orchestrated frigidity created by Erik Satie's furniture music.

FURNITURE MUSIC

Coined in 1917 by French composer Satie, *musique d'ameublement* can be more literally and accurately translated—in terms of its waveformed intent— as 'furnishing music'. Just as seating in a room might be specifically placed to encourage social interaction, Satie's music was situated in a location's background to perform a similar task. Although regularly cited as an early precursor to Muzak (because of its utilitarian sonic provision), it would be a mistake to conflate furniture music with the affects or effects of stimulus-progression programs. Indeed, Satie's musical proposition had a very different relationship to the psychological and physiological outcomes that it tried to inspire in those within its reach. Whereas Muzak in the Fordist factory aimed to help choreograph the body into the repetitive routines sequenced by its new mechanical counterparts, furniture music pursued an antithetical agenda: Those within earshot were asked to completely ignore it rather than attempt to move to or follow its rhythms and cadences. A virtual manifesto of disengagement, Satie's radical sonic gesture, with its dadaist inclinations, must have left many 'listeners' perplexed. The notion of human social interaction being encouraged by the composition's seemingly benign presence in a room was without precedent. Operating just beneath the threshold of intrusion, music would be used to gently oscillate communication by embedding melodic threads into the texture of discourse, so that when verbal exchanges became stilted, Satie's arrangements were there to help suture the silence.

According to Joseph Lanza, there is a direct correlation between the advent of a form of music that would furnish commodity-based voyeurism and the emergence, during the same era, of a new type of city dweller. As a vaguely arcane prism of consumerism, the shop window acts by telescoping desire. The musical accompaniment to this nascent de- and refocusing of corralled aspiration was therefore an inevitable development for Satie, given that music had always furnished all manner of social and private activities, from weddings to ritual killings, from waiting to shopping. Making clear the context that this new kind of sonic fitting would adorn, Lanza writes, 'Various social and economic forces during Satie's time imply that he did not arrive at furniture music out of an aesthetic bubble. Mid-nineteenth-century Paris was among the first European cities to shift from a market economy to a consumer culture. The Parisian department store, for instance, replaced the quaint, small shop. The Bon Marché appeared by 1870 to transform shoppers into ambulatory voyeurs propelled by a fixed-price system, rendering haggling obsolete and gawking all the more enticing' (Lanza 2004, 20).

Contemporary modes of gawking have been upgraded and now have a wider impact, given that consumer interactions are now catalogued and analysed for predictive behavioural modelling. With the addition of flat-panel speakers, the gaze is reeled in by sonic anchors whose efficacy owes to their absence of depth, devices that leverage their hold on these sheer surfaces of reflection. Engaged in the act of multisensory seduction, they cut through the static dynamics of monosensory perception. Akin to the Whispering Window's intention to be invisible, Satie's objective with furniture music was for the mode of reproduction—the players—to dissolve into the background. If the premise of autochthony is that of being indigenous or native to an inhabited environment, then we might say that this music was aiming to be *sonochthonic*: it sought to persuade one that it had always been there—not just an aural accompaniment to the fixtures and fittings, but *part of the furniture*.

'Performed' during the interval of Max Jacob's play *Ruffian toujours, tru- and jamais*, which was staged at the Galerie Barbazanges in Paris, Satie's *musique d'ameublement* was played by a (dis)located quintet situated in isolated positions around the space—anywhere other than a structure that resembled a stage. As noted by Caroline Potter, 'The music had no visual focus to help dissipate musical attention' (Potter 2016, 112; first published in 1952). This did not stop the general public from watching, however, even when Satie implored them to not listen but, rather, move around the space as 'normal' (Milhaud 1952, 153). Insisting that one not listen was, and is, pivotal to the operative mandate of both furniture music and Whispering Music. But while the latter, unlike the former, wishes to attract attention, Whispering Music has no interest in maintaining aural concentration over any sustained period or in any idea of listening that is antithetical to the store's economic remit to get customers through the door and absorbed in other types of transactions. Through both of these waveformed processes, organised sound, noise and silence map onto each other in ways that challenge the validity of their discrete classification. In the case of Satie's furniture music, this fluidity posed questions that are still argued about today.

Satie implicitly understood the more abstract consequences of othering music—of producing music that was nonmusic. Employing an ocular metaphor allows us to come at this from a different perspective: by telescoping this othering logic out, music can be interpreted as sound; twist back the resounding lens and it can be comprehended as environmental texture; the final turn brings into focus an elementally obtuse and wider picture—music as a vibratory potential. With this latter distinction in mind, after a bout of initial disappointment at the public's voyeuristic response to his nonmusic, Satie wrote bullishly to his friend, cultural polymath Jean Cocteau, declaring, 'We want to establish a music designed to satisfy "useful" needs. Art has no place in such needs. Furniture music creates a vibration; it has no other goal; it fills the same role as light and heat—as comfort in every form. . . . Furniture music for law offices, banks, etc. No marriage ceremony complete without furniture music' (quoted in Shattuck 1967, 27). Satie's provocation—'Furniture music creates a vibration; it has no other goal'—was a boldly futuristic statement to make in 1920s Paris, a sonic fiction that would only find its scientific realisation nearly a century later.

Waveformed expressions proposed to be perceived and utilised purely as vibration, rather than as music or nonmusic, are rare outside of the discourses and frequency-based practices of religions or pursuits with spiritual connotations and/or aspirations. One exemplar of such ritualised vibrational activity would be the singing bowls used in Buddhist and Taoist meditation and chanting. A mallet is wielded and turned consistently around the outside rim of the bowl, producing a sustained note. This drone texture subsequently becomes the sonified environment in which forms of meditation and relaxation take place, listeners supposedly clearing their minds and bodies of all thoughts so that a direct personal channel with the instrument opens. It could be argued that overarching Buddhist metanarratives pertaining to emptiness, along with the observation that there is a lack of intrinsic meaning in any perceivable object, are apposite to Satie's compositions (themselves emptied of any distinguishable or 'thoughtful' content), but this would miss the point. Furniture music does not attempt to connect all matter in a voided equivalency, as does Buddhist meditation. Instead, furniture music functions by emptying itself of all connection with those within hearing range so that oral exchange can become communicable in the relational vacuum thus opened up.

In many ways, this line of reasoning appears to deliver us back to thinking about the functionality of Muzak and its ordering of space, movement and presence through rhythm and tonality. And yet, while furniture music is not necessarily the antithesis of Muzak, it does diverge from the latter in terms of both process and outcome. While Muzak was originally deployed to choreograph the industrialised body into new working relations with machinery, Satie's compositions were supposed to pass unnoticed and therefore to be passive to the point that other types of behaviours and movements would negate any perception of them at all. As such, the compositional motivation was its own negation—a sublime act of aural self-immolation, an obsolescence that could only be truly rendered viable by the emergence of a more dominant form of sonic ecology. This is sound deliberately bred to be uninspiring, insipid and debilitated—the antithesis of the Darwinian evolution that has governed much of the past century's organised sound, which has evolved and propagated itself by competing on the grounds of seductive, attention-grabbing intensity.

MUSIC AS COMPETITION

Forms of music that can be identified as being comprised of competitive elements or being instigated by competitive behaviours are often related to the modes of movement that they inspire or are inspired by. Musical and dance genres such as capoeira, rap and breakdance, footwork, and house and voguing are just a few examples of musical and movement-based dynamics that feed each other's growth and shifts in production and performance. A truncated lineage of one or two forms of dance and music is hardly evidence of widespread cultural competition, but the globalised obsession with televised musical rivalry is more so. *Star Academy Arab World* (Lebanon), *The*

X Factor (the United Kingdom), *Kōhaku* (Japan) and *The Voice* (the United States) are but a few of the hundreds of hugely popular shows in which music is used to pit individuals, groups and audiences against each other. This interactive and child-friendly weaponisation of culture appeals to the mainstream because it lures the listener and watcher into the fray of competition and the feeling that, even if they are simply voting, or just tuned in, they are somehow involved in the production of meaning derived from conflict.

The most conspicuous incentive within this process is the opportunity to derive clear-cut and obvious decisions—one of the reasons why competition, in all forms, has become so popular. It delivers us out of the complex and endlessly reverbatory micropolitics of everyday decisions that are more intensively recorded and whose ramifications are more easily traceable than ever before. The butterfly effect, in which micro events can scale up to translate into macro outcomes, can be seen at work in everything from social media to dataset analysis, but it is the capacity to capture the processes in granular detail that immerses everyday living into a palimpsest of social webs and entanglements. By endowing an activity, an utterance or a song with a clear-cut distillation of meaning, competition removes the midrange, delivering only high or low, silence or noise. As such, competition becomes a quest for escapism, a removal from the Bitcrush density of information that envelops our every decision to a stage on which the duelling tyranny of zeros and ones plays out.

Satie's furniture music is captivating partly because it eschews any involvement with the language of competition. It questions the archetypal tropes around the production, distribution and reception of music that render it such a powerful tool in terms of the production of identity, the cohesion of social groupings, sexual seduction and even physical violence. It was a music that was out of time by design. Considered in these terms, the premise of furniture music and later associated musical genres, such as ambient, new age and vapourwave, stand in even starker contrast, both in terms of their acoustic blueprint and, more importantly, their mode of functionality—or, given these genres' noticeable recalibration of sonic efficacy, their aura of deliberate and adversarial dysfunctionality.

TEFLON ACOUSTICS

In the face of the Darwinian evolution of popular musical forms, Satie's furniture music becomes the poster child of unnatural selection, a musical metaphor for the survival of the enervated. The type of selection promoted by the Whispering Window is, on the surface, a more mundane reflection. But then it is not the promise of consumer choice that is in question here; rather, it is the sonified process that entices the transient to become momentarily static, at least long enough to register the glazed aspirational materiality that, like the window, oscillates in a constant state of hushed anticipation. Analogous to the mode of self-induced devitalisation transmitted by furniture music, the Whispering Window finds its industry in lo-fi emission rather than in the accuracy of acoustic reproduction aimed at by hi-fi technologies. Eschewing high fidelity and separation in favour of compressed ethereal pitches that scratch their way across the clear divide that is the window, the disordering parallels between the two come into conversation with each other again: a century-old message in a bottle.

From a 1920s intermission, this pitch(er) of sonic data has been heated by the fires of capitalism and rolled out into the upright event horizon of a window that holds light and sound within its frame. Read through an alternative prism, Satie's message comes to make more sense in an era of contemporary capitalism, when traditional divisions between labour and leisure have been eroded. Methods of music production, distribution and consumption, meanwhile, reflect the surface culture of Teflon acoustics, in which nothing sticks and indeed nothing is meant to. Overload capital requires such transient sonic channels in order to flow through the flattened vista of flattened speakers. This is an indexed temporality in which one is considered most financially viable when in automated leisure mode, spending and redistributing the currencies of recreation, which in turn justify the cryptorationalisations of the free-time economy.

From this perspective, Whispering Windows are merely the latest iteration of a stealth economics that deals in notes of a musical variety rather than those endowed with an exchange value. Thus the lineage that runs from furniture music through Muzak, ambient and vapourwave connects in turn to this recent transmission mode that functions in alignment with capitalism and its fluid dynamics of disposability and inbuilt obsolescence. Systems of (musical) competition are largely energy inefficient, the friction of contestation slowing down and obstructing the flow of information and products. The sheer surface quality of furniture music and whispering music with their benign don't-mind-me manner of seduction reflects a will to enter the materiality of everyday living without asking permission, weaving together experiences, no matter how divergent, into a seamless flow of consumption.

The coalescing of transmission techniques with modes of listening and production aesthetics underpins an emergent sonic culture that deviates from the competitive acoustic dynamics discussed above. These more recently occurring listening practices are subsumed by the pleasure of instantly forgetting, of being sustained by sonic textures rather than getting bogged down in the personalised narratives and identities conveyed by music that is endlessly, aggressively, competing to be different. This is precisely why streaming services such as Spotify are successful: their bespoke tailoring of musical textures for users who are busy filling their time with other work or leisure pursuits is their main promise and attraction. With such services, the consumer does not need to make choices, marking them as vibratory tendencies in perfect alignment with contemporary capitalism. Decision-making slows down rates of consumption, whereas being tightly woven into the very fabric of consumption by the sounds of acquiescence is good for business, however you listen to it. In this analysis, paradoxically, textures of average fidelity become the apex predators of the aural environment.

In the world of home hi-fi equipment (the acoustic antithesis of the Whispering Window), pressure has been exerted on audiophiles over the past fifty to sixty years to own exceptional technology that delivers an undistorted reproduction of music with frequency ranges pristinely separated and fidelity maintained at high volumes. With the transfer of technological desire from home systems to mobile devices, younger generations demand new affordances with diminished sonic capacity. With such mobile technology in mind, it is fitting that our furniture or our surroundings speak to us in a lo-fi manner. We live in an age of hypercompression, where traditional notions of high quality are regularly forfeited for mobility, transfer speed, (wireless) connectivity and ubiquity. This is not to disparage the prevalence of compression, which has been driven by the culture of digital production, distribution and archiving; whether it is music, film or photographs, the majority of us now engage with these mediums via compressed formats such as MP3, MP4 and JPEG. We are willing to forego higher-quality formats for the immediacy of being able to download and transfer content to other mobile technologies so that we can consume them on the go—walking, in the train, in the car or on an aircraft. In this, light flat-panel speakers make perfect sense for the transitory times we live in, for they are the most mobile of sound systems.

Given the ubiquitous nature of flat surfaces within the urban milieu, Whispering Windows (or future forms of the technology) promise to be the most pervasive disseminators of surround sound. Mirroring the compressed content that literally envelops and punctuates our every step, journey and fragment of downtime, the resonating surfaces of Whispering Windows' tempered glass, with their compressed surfaces, assert a new cartography of pressure points upon the body—a range of forces that travel beyond the dermal interface. As previously suggested, compression is a type of formatting that extends across all types of digital media. The technologies that enable and support its proliferation are constantly at our side, in our ears and in front of our eyes, on our arms and around our wrists, in our heads and hands. With technological augmentation being so omnipresent and the fluidity of compressed content allowing us to flow in a seamless tide of multisensory narratives, the concept of any surface potentially becoming a speaker that can interface with such devices is no longer a great leap. And as more objects—whether phones, fridges or cars—speak back to us via AI-enhanced chatbots, it is, in many ways, a predictable trajectory.

Increasingly, sensors, cameras and wireless devices are being integrated into architectural and landscape design. Data centres are proliferating, siphoning off data from 'smart' cities and redistributing the datasets being accrued. Given this irrevocable drive to record, reprocess and analyse, it would appear that the human population of these urban environments, in order to benefit from this 'smartness', will need to acquiesce to their presence being responded to rather than merely reflected. With information pertaining to our every footstep, purchase, bus ride and pollution level finding its way into a data set, the urban environment is increasingly becoming aware of its inhabitants' movements, desires and shortcomings.

Speculations about the city becoming an entity that supports distributed intelligence and agency are becoming a regular part of discourses around the built environment from a range of perspectives—architectural, fictional, technological, human rights and economic, to name but a few. If and when this sentient city comes into being, a new approach will be required that canvasses all of those disciplines to formulate a robust set of strategies. They will be tasked with renegotiating our spatial and temporal relationships with this newly evolving sensory climate in which all of our somatic mechanisms for collecting information have been massively extended and new ones added, and yet the data accrued exists at one remove from our internal processing facilities. Our skills in negotiating datascapes will increasingly become more vital than those required for navigating physical terrains. Questions such as those posited by digital-humanities theorist Caroline Bassett echo such a sentiment: 'How far does the mode of perception within which the mobile operates relate to the way we prioritise mobile space over physical space?' (Bassett 2003, 348).

To further process what this might mean, a thought experiment would be instructive here. Let's project forwards (or diagonally) a little, to an era in which Whispering Windows live up to their insidious name. Bypassing human presence and agency altogether, they communicate with our mobile devices rather than needing to interface with their comparatively slovenly carriers. Mobile space will be formatted via convergences of digital and concrete realms. Their intersection and integration will be superfluid in nature, with such union ensuring that connection, analysis, prediction and action take place within the time it takes to set up a selfie. Within this intensely linked and looped municipality, the significance of the way in which humans travel through space will increasingly be predicated on the dynamics that govern our devices' connection to their environment.

Let's extend the speculation and set a scene: We are on a busy but quiet twenty-first-century smart street, where a driverless taxi has just pulled up and hydraulically lowered its floor to the exact level of the pavement. Out steps a retrofitted flâneur. Her converged mobile device makes contact with the window of a holoentertainment centre she has been dropped off in front of, causing personalised adverts to be triggered and aimed her way. Since her presence adds to the number of people active in the area, the device also contacts local health services, helping hospitals and clinics make predictions for the night's patient load based on gathered data on all of the locality's active inhabitants. The mobile has also spoken with the solar panels embedded into the pavement, a signpost thirty metres into the distance and a drone flying overhead delivering encrypted packets to a data analyst in the suburbs. All of this before her feet have even hit the ground.

In light of this connectivity, it is quite plausible to foresee adverts being delivered in a personalised manner, tailored to the profile of the person who has booked a taxi or a train ticket, for example. All transferable data in this scenario would be tailored to the traveller's likes and dislikes on social media, their consumer profile, credit history, political allegiances, travel itineraries and purchase cosmology. A forensic model of the oscillating databody is being drawn up, distributed and fed into a new mapping procedure that reimagines urban space.

Urbcom bots arrange and orchestrate all communications data extracted from our city's constant exchanges between the traditionally understood structures of nonmoving buildings and the fluid architecture of its vehicles and mobile devices, from the minutiae of our flaneur's taxi exit to the large-scale metanarratives extracted from environmental data sets. Like spiders patrolling a web, able to perceive the exact location of the slightest vibration of any creature or object that comes into contact with it, the bots are agile and powerful enough to be able to pinpoint specific communications if required. Meanwhile, the augmented silence within which our wandering soul manoeuvres fills a relatively impoverished bandwidth known as the human hearing range. Outside of this, the city is noisier and more densely stratified with frequencies than ever. The paradigm of listening and conversing has been shifted to the echelons of the unsound, in which the human voice ranks alongside a thousand other transmission variants, and below the majority of them in terms of speed and efficiency.

Connection rules. Everything is in conversation with everything. Animism has found its materiality in the silicate dependence of globalised progress.

All space is mobile, and humans have become the equivalent of pack mules, carrying mobile nodes of the coms system around with them. As they slowly route their ways through the roads and architectures that support this massive, diffused nervous system, somatic soft/hardware has become the embodiment of the uncompressed, too unwieldy and poorly formatted to be able to flow with the waveformed emissions that envelop it, struggling to carry out three or four, never mind a dozen, forms of communication simultaneously. Our highly prized and seductive twenty-first-century figure of the hypermobile subject has been outmanoeuvred by the very terrain through which it used to move. The rationale of presiding over all that one can see has been usurped by the flowing potential of the vibrational intersectionality of all that one can hear (but can no longer necessarily speak back to).

In this scenario, human agency is subservient to that of the built environment's networked intelligence. While we might not find this a desirable trajectory, there can be little doubt that the majority of urban planners, developers and inhabitants want our cities to take care of us (while acknowledging that this implies a need to track, influence and demand). This is evidenced by our will for cities to be environmentally sustainable, to make decisions for us (by predicting our needs in real time) and to speak to us (via voice-activated chatbots) when we demand it. Such a programmed drive to conserve humans (for now) is also shaped by the expectation that economic growth will be generated from urban agglomerations. Of most interest here, however, are the thresholds between collaboration, protection and domination and how we are to interface with the technologies that offer all of these dynamic adjacencies.

The future projected above only enters the realms of dystopia when the communicational palimpsests of which humans are a part begin to treat their presence as a hindrance to an otherwise fluid materiality. Communications networks and systems demand data formats that can quickly and easily shape-shift. Future cities will stipulate that humans do the same in relation to the way the built environment reveals itself to those navigating it. How we sensorially decompress and expand as we navigate this constant displacement will be crucial to comprehending its ever-changing coordinates. Such expansion will help determine our relations to space, presence and movement as we compete for portable coherence. No longer will attention, and by extension presence, be a matter of prioritising; rather, it will be a case of our ability to transition between diminished, augmented, virtual and expanded realities. To speed up and become fluid or expand and hold space when necessary. To compress or not to compress.

In the realm of information technology, lossy compression reduces bits by identifying unnecessary information and removing it. Behind this seemingly benign technological process resides an array of decisions that have considerable racial, social and economic ramifications. For who exactly is it that decides what information is excess to requirements and why? Putting sonic-compression formats under scrutiny, Matthew Fuller declares that the development and proliferation of certain musical formats reveals the influence of Western technonarratives steeped in racial bias. The embedding of frequency-based prejudices into the technologies that soundtrack our daily lives begs the question of all the other forms of data that modulate our daily experiences and decisions. For Fuller, the shift in transmission and listening habits in the West comes loaded with genealogical and biological intent. He writes, 'The MP3 file format, which has achieved such mass usage as a means of circulating tracks via the Internet, is designed simply to match the included middle of the audio spectrum audible to the human ear. Thus it obliterates the range of musics designed to be heard with the remainder of the body via bass. This is not simply a white technological cleansing of black music but the configuration of organs, a call to order for the gut, the arse, to stop vibrating and leave the serious work of signal processing to the head' (Fuller 2005, 40).

Challenging dominant models of compression and transmission, Fuller's arguments about the ways in which we are unconsciously obliged to listen and not listen are germane to the notion of decoupling attention from environmental stimuli. The modality of not listening here, however, is reversed in terms of decision-making processes; algorithms make the choice for us so that a tacit habit of not listening to that which is deemed to be surplus to requirements is directly embedded into the listening activity itself. And it is precisely this surplus that is removed in the act of compression, from the music that furnishes and from the windows that whisper. In this triangulation of unhearing, R. Murray Schafer's (1993, 77) premise regarding sonic imperialism's causal relationship to the domination of space via loudspeakers needs to be rethought. Unhearing is a hushed negation of his maxim, 'We hear sound. We belong to sound. We obey sound' (30). The loudspeaker has been exposed as an orbit of spent resonance. It is the time of the quieted.

The period between 1900 and 1940 was quite the opposite. Charting an evolution of attitudes towards noise between these periods, particularly in relation to European and North American noise-abatement campaigns, historian Karin Bijsterveld gives us an insight into the fluctuating narratives and symbolism pertinent to that forty-year period. Her observations highlight why noise, and by extension sound in general, is welcomed and reviled in equal measure depending on the acoustic prism through which it is perceived. She writes, 'Making noise was still seen as a sign of being uncivilized, of having no manners. In a more general sense, noise was thought to have been welcomed because of the positive connotations, such as progress and power, that loud sounds possessed' (Bijsterveld 2003, 179).

As discussed in chapter 1, it is within this same context of the factory that silence came to represent an expensive glitch in production. Machinery would only stop during working hours if there were technical problems that brought the assembly lines to a halt. Today this fear of silence has been redefined once more, as it is perceived through a new technological schema. While musical compression attempts to remove noise and excessive frequencies from all genres of music, adjacent conventions for the industrial compression of the sounds made by products such as cars, fridges and computers are also being negotiated. Silence no longer designates failure but stealth.

BONE LOGIC

The above claim already applies to HSS technology's capacity to silently track individuals without announcing itself, evidencing its military provenance and its possible future. But as much as it might seem to make sense to align Whispering Window technology with this same sensibility, a divergent approach comes to the fore via the model of the flat-panel speakers fitted onto train windows. Originally developed for Sky Deutschland broadcasting service in 2013 by the German advertising agency BBDO, the Talking Window fundamentally changed the transmission mode of the flat-panel speaker by using the technique of bone conduction. As the vibration of the train on the track lulls the passenger to sleep, a more focused and formatted vibratory procedure inserts adapted adware into the cranium as messages are delivered into the skull of the tired traveller who has let her head rest against the window. In this scenario, the adage that the medium is the message takes one step closer to being literally true: bone logic.

This iteration of the flat-panel sound system quietly insists on a more intimate relation to its waveformed agenda, one that receives and transmits without cognitive intervention. In doing so, it renders the most fully formed, and most troubling, composition of the modulating antenna-body we have examined thus far. While the storefront flat-panel system and the HSS both transmit frequencies that, even if surreptitious, are audible to the conscious mind, the Talking Windows fitted on trains take a different approach altogether, one involving a way of hearing that has yet to be discussed—bone conduction. In our conventional understanding of hearing, molecular disturbances in the air that are within the human hearing range of twenty to twenty thousand hertz are picked up by the eardrum. The eardrum decodes or translates these fluctuations into a new set of vibrations that are then sent to the cochlea or inner ear. Bone conduction works instead by conducting sound to the inner ear via bones of the skull, circumnavigating the outer ear and the eardrum altogether.

It works so effectively because the cochlea is itself a conical chamber of bone. This provides a material explanation as to how vibrational information is carried from the outer skull into the organ of Corti, the sensory organ of hearing that sits within the cochlea.

Bone conduction can be split into two transmission processes—one that relates to higher frequencies, the other to lower frequencies. Compression bone conduction refers to the higher-range of frequencies that induce certain sections of the skull to vibrate, the direct compression of the otic capsule (the bony ring that surrounds the inner ear) causing the oscillations to be transmitted to the cochlear fluids. Inertial bone conduction, meanwhile, explains how the lower-range of frequencies, below 1,500 hertz, affect the skull, which resonates as a rigid body. As stated by Dr. Tobias Reichenbach, from the Department of Bioengineering at Imperial College London, bone conducts lower-frequency vibrations better than air does (Colin Smith 2014), which explains why we perceive our own voices to be lower-pitched than they actually are, an illusion liable to dissolve when we listen back to recordings of ourselves. Our vocal articulations are thus augmented with a phantom bass register that exists because of the bone channel that transmits directly to the cochlea. So when a tired commuter, for example, sits down and rests her head against the Talking Window in order to sleep, she is also enacting another process at the same time: the contact of bone against glass completes the circuit that allows marketing information to be transferred directly into her head.

What is of particular note here is that this technique aims to interface with the mind when it is in a sub- or preconscious state. Bypassing the editing suite of consciousness, Talking Windows fashion their efficacy around a body with a decreased ability to react to stimuli. Picking up on one of the more expansive threads of this chapter, it may be observed that flat-panel-speaker technology distances the notion of the audience from the notion of active listening. And beyond this, flat-panel-speaker technology challenges our preconceptions as to what a listener is. The body can exist in a range of perceptual states, from that of heightened or expanded awareness to that of sleep, from synaesthesia to coma, along with a huge number of other states that are more elusive when attempting to qualify or name them. The body is in constant shuffle mode in terms of the way it receives, translates and emits frequency-based information.

Each condition implies a different relationship to the perception of internal and external stimuli, the scale being broader for hearing because we cannot shut it off in the same way that we can shut off visual information. In this series of contextual relationships, the listener encompasses a range of wave-formed subjectivities, each one with a different focus, attention and capacity according to the circumstances. Talking Windows transmit most successfully when a human is in a state between rest and sleep, with aural information bypassing the primary mode of consciousness and the associated matrices of decision-making processes that accompany it. Deviating from traditional notions of listening, Talking Windows cranially creep into our imaginations.

Differing from infrasound inasmuch as it can work with frequencies that span the audible and inaudible spectrums, bone conduction could be considered an unsound technique, as it bypasses the outer ear to deliver its vibrational payload. Its mode of delivery differentiates it from the aural subterfuge practiced by the Ghost Army, for example, as bone conduction is more concerned with a recalibration of both the sensorium and the spatial. As with other techniques, content and technologies discussed in previous chapters, the modality of the Talking Window is that of the virus as it enables waveformed parasites to enter the body unnoticed. In this analysis, bone and glass fuse into an oscillating topography that the sound traverses. No material distinction is made by the infecting agent, as categorisation between living and nonliving things becomes obsolete, oscillation rendering all material objects simply as channels through which frequencies pass.

In the frame of the Talking Window, bone equates to train equates to landscape in a schema that fundamentally apprehends all matter as mobile space, the variation in its density and its capacity to transmit vibration being the primary means of differentiation. Translating this notion back into a human-centred perspective, information transfer becomes indelibly linked with the hyperconnected digital space that affords the era of the *Internet of Things*—or IoT. Commenting on the period preceding the advent of IoT, and in particular the multiplicity of mobile spaces that technologies allow and encourage humans to move through, Bassett writes that 'an older sense of the distinction between the landscape and the journey, and of the spatial dynamics underpinning this distinction, no longer pertains. These days, as mobile-equipped travellers, we operate in that speed-blurred band that used to demarcate the division between landscape beyond the rails and the fast-moving space of the train. Or, rather, there is no longer a boundary but only an *interface*. You are advised to "take your world with you" when you go because this is the end of the incarceration vacation with its unexpected freedom and its constraint. The question is what comes next' (Bassett 2003, 346).

Answering this question back in 2003, Bassett was correct in citing the erasure of the boundary and the proliferation of interfaces as the markers of change in the composition of emotional topographies and transgression. What has changed, however, is the distributed location of, and the agency imbued within, the interface. It is no longer solely bound to the consumer-friendly destination of the pocket-sized mobile device, as myriad portals to the fluid digitality of the Internet are now located in a disparate assemblage of objects, from doorbells and juicers to traffic lights. As a waveformed version of the Internet's promise to render all in binary, the potentially omnipresent terrain of the flat-panel speaker anticipates a future sonic ecology, a frequency-based cartography in which everything communicates.

Within the changing aural ecosystem described in this book, the Talking Window once again alters the perspective and spatial relations within which sound systems find their efficacy. The flat-panel speaker negates the HSS's scopic targeting of the individual, rerouting its epistemic shift by slowly recoiling back and only becoming effective when body and glass become one. The Talking Window sound system no longer surrounds nor does it target. Instead, it converges with a body that becomes an element and extension of it. All traditional notions of the speaker are obsolesced by this problematising of the intersection of space, transmission and cognition. The technology only functions insofar as it loses all of the qualities that have come to define a speaker as such, from the physical design of the diaphragm or coil to the expectations related to air movement. Effectively, the speaker has been dematerialised, its fetish value and status dissolved and dispersed into a fluid agency of objects—a dispersed delivery system for the only force that can lay any kind of claim to universal agency: vibration.

Mapping out a sensory history of the eighteenth and nineteenth centuries via vibration, Shelley Trower examines the anatomical logic of the oscillating body in relation to the resonant environment it is part of. Supplanting nerve with bone, Trower supplies us with a historical precedent for the flat-panel speaker in the form of a musical technology that similarly tracks the orchestration of shifting relations between all objects. She writes, 'The Aeolian harp provides an earlier, acoustic model of embodied consciousness, which could serve as a bridge between the "classical" and "modern" accounts of sensitivity. This model presents sensations as originating in *both* an external stimulus and the body, especially the nervous system' (Trower 2012, 15; emphasis original).

The nervous system at play in the expanded speaker network suggests connection—of the sensorial body, of all material things—and yet it also connotes isolation in its all-hearing, all-feeling capacity to strategically target the individual and to separate it, both when showing signs of life and when dead to the world. In this sense, the antenna-body finds its most germane aural context in the relation between the passenger's body and the train, both vehicles with stimuli passing through them and projecting back out. Describing mid-nineteenth-century attitudes towards the oscillating humanoid, Trower continues, 'Hearing and speaking, reading and writing, the reception and radiation of energy both within and out of the brain was conceived as one and the same process. Not only were people, described as mechanisms, understood to be sensitive to the various forms of vibratory energy, but energy was conceived as the driving force behind life and consciousness itself. This is a body without borders . . . simultaneously receiving vibrations and radiating outwards, while it becomes impossible to distinguish between inner and outer reality' (ibid., 48).

A historical version of the antenna-body can be heard loud and clear in this description. As transmitted throughout this text, the subject that Trower describes being listened to and felt is one that is simultaneously conduit, channel and agent. While reconfiguring the material efficacy of the somatic, this perspective also resituates the body both in terms of narrative sensibility (it is decentred, no longer the pivotal element) and in terms of its relation to volume (spatially and sonically). The physical volumes of receiver and transmitter need to be aligned in order for the waveformed volume to become perceptible. This gap that had always existed between the ear and the diaphragm is now as obstructive as the ear is limiting. Its lacuna represents a loss of vibratory coupling—a dislocation in the measure of the acoustic compass.

The storefront model of Whispering Windows has more in common with the HSS in terms of a discrete contouring of space. But it is the Talking Window on the train that radically alters the geographic relation and disposition of the speaker system. If the primary dynamic of a traditional speaker system—to move air in space—is cut up by the HSS, the very need for space is cut down by the premise of the Talking Window. Distance and its precise demarcation (along with all the requisite audiophile obsessions) is the antithesis of what is necessary to achieve the ultimate experience of this sound system. In fact, transmission can only be achieved by the collapsing of such concerns.

As Steve Goodman points out, 'When any surface becomes an emitter of sound, the distinction between loudspeaker and environment further dissolves' (Goodman 2014, 169). In the world of the Talking Window there is no time or space for the sweet spot in the room. The listener's position, rather than being dictated by geometric detachment is now made relational by its connection to the environment it exists within and its capacity to further extend the voice that passes into and through it. Bone converges with the machine converges with bone. Augmenting the role of the conductor within the aural ecosystem, our train passenger is a softly mechanised part of the flow, a vibratory adaptation of Gilles Deleuze's body without organs.

HAUNTED BY CAPACITY

One of the most significant narratives daisy-chained throughout the chapters of this book concerns the ways in which speaker systems have progressively been designed or repurposed over the past century to isolate the body. The bone-conduction technique used on the train's Talking Window is incredibly intimate and detaching. An ossified quarantine. And yet this is not a manifest confinement. For the sleeper does not necessarily know they are interfacing with anything at all, much less a transmission technology. With the insertion of sound systems such as earbuds into the body, a tacit intimacy is achieved that transports the carrier into other auditory and imaginary orbits. When that mode of transmission is flattened out in a parallel ontology that releases such intimacy from personal choice, leisure and pleasure, a whole new plane of potential utopian and dystopian futures slides open.

In chapter 3 it was argued that the technique of heavy radio rotation can be understood to have reached the end of its logical track in the dystopian cells of Guantánamo Bay and Abu Ghraib, and so the conception of the acousmatic is propelled to its final destination here. The Talking Window rearranges the composition of source, transmission and listener so that the space that once separated activity and object is collapsed, any echo of the omnidirectional composition of quotidian urban acoustics dampened by the direct somatic bridging of the waveformed experience.

The original French term *acousmatique*, coined by composer Pierre Schaeffer, can be traced back to the name Pythagoras gave students who had recently joined his circle of acolytes. They were known as *akousmatikoi*, or 'hearers'. On the train, the mythical wall or veil from behind which Pythagoras supposedly taught these students (so that they would concentrate more deeply) has lost its opaque qualities and become transparent, while the requirement for committed attention has been replaced by a preference for receptive somnolence. The speaker has left the building, the train and any other kind of structure or box you might want to put it in.

As previously noted, the remit of the Talking Window or Whispering Window is not to dominate acoustic space in the manner of the boxed stereo or surround-sound systems explored in the previous chapters. Nor is it to function in the manner of other sound systems that existed before electricity and that had similar objectives but fewer energy resources. Instructive here is Alain Corbin's research into earlier forms of public sound systems, which were cast in copper and tin. He describes how bells were employed in nineteenth-century French villages to sonically denote spatial territory, to maintain the religious symbolic order, to share information about social events and to organise collective temporality (Corbin 1998). These bells were prescient forms of sonic public address because they denote a pervasive tendency in our thinking about sound's ability to demarcate space and, by extension, territory—and when we utter the term *territory* we implicitly speak of those presences that should be included and excluded within it, along with the sedimented raft of political and socioeconomic structures that inform such decisions.

Corbin declares that 'a bell was supposed to be audible everywhere within the bounds of specified territory' (ibid., 120). Describing the power of this heavy sound system to organise space, Corbin goes on to describe how 'the range of a bell, inscribed in a classical perspective of harmony, served to define a territory that was haunted by the notion of limits as well as the threat of those limits being transgressed' (118). All sound systems, from the bell to the flat panel, encompass—for those within hearing and feeling range—this dualistic tendency to internalise and externalise, summon and purge, seduce and repel. The limits of surfaces that emit capitalism's spectral archive of influence are possibly less bounded than ever before, if for no other reason than that which we used to call the limit has now become instead a point of connection. The boundary has become haunted by its capacity to be fused with other waveforming phenomena.

STONE TAPE THEORIES OF ANIMISM

A history of objects imbued with agency and, therefore, potentially, voices, would include many and varied examples. An obvious place to start would be with animism, a religious belief of countless indigenous peoples, proposing that all material phenomena possess agency and some notion of sentience. Long before Cartesian dualism gained any philosophical traction, there existed dominant ways of thought that did not place boundaries upon the spiritual and the material. As such, many early forms of animism, such as those imbued within Native American religions, Shinto (Japan) or Korean shamanism, promoted the belief that rocks, trees or water have a spiritual essence or life. Contemporary writers and thinkers, meanwhile, such as Iranian philosopher Reza Negarestani, speculate on animistic potentials in books like his *Cyclonopedia: Complicity with Anonymous Materials* (2008), a theory-fiction text postulating that we live in a world in which oil has a sentient capacity and has historically controlled human interaction with it and the destiny of the world at large. Commenting on the role played by Whispering and Talking Windows in bridging worlds, Goodman points out that 'the science-fiction future that constitutes the present of animistic capitalism appears to rejuvenate ancient civilizations that believed that spirits inhabited objects. Sound projected through any object whatsoever injects quasilife into the inanimate' (Goodman 2014, 170).

This assertion that sound passes through a material object to create other manifestations of existence could also be applied to the quasianimate, the nonmaterial. Holographic projections of dead rappers for concertgoers—*rapparitions*—would be a prescient place to start. Also referred to as 'original virtual performances', here the camera essentially captures the spirit of the performer it has recorded; it is archived in a digital ledger and summoned back to life by a silicate alchemy comprised of sonic nostalgia, rotoscoping technologies and expanding Lazarian economies. But this is not, in essence, a modern industry, not by any stretch of the imagination: Giambattista della Porta first conceived of it in 1558, describing an illusion, 'How we may see in a Chamber things that are not' in his work of popular science *Magiae Naturalis* ('Natural Magic'). Three hundred years had passed by the time John Pepper and Henry Dircks manifested della Porta's idea in (virtual) reality by making a ghost appear on stage in Charles Dickens's theatrical rendition of *The Haunted Man* in 1860 (as described in Steinmeyer 1999).

On stage in the twenty-first century, it is difficult to tell the difference between the living and the dead. In 2012 the Digital Domain Media Group revivified rapper Tupac so he could play live from the grave alongside Snoop Dogg (who claimed that the encounter was 'spiritual') and Dr Dre at the Coachella music festival in California. In his own inimitable way, Tupac intoned to the audience, 'To lead the wild into the ways of the man / Follow me; eat my flesh, flesh and my flesh' ('Hail Mary', in Shakur 1997): a zombie-call for future bloods to become immortalised by digital divinities—a waveformed flow of speculative vitality. Here frequencies delegate an invested mortality that oscillates between augmented, virtual and diminished realities. The future figures of the body (and the income that will be accrued from them) animate an amortised economy in which 'not only the labor but the laborer himself has been rendered immaterial, conjured up, and put to work. Outsourcing here takes on the character of "outsorcery", a conjuring of the dead to do work once the sole province of the living' (Freeman 2016).

A connected but more materially bound locus from which to explore contemporary sonic archiving and projection is afforded by *Stone Tape theory*. The precursory writings by philosopher of perception Henry Habberley Price in 1940 suggest the formation of 'place memories', where memories lost by a human host are archived by an environment such as a house. In this tormented equation the stored information can later be accessed or experienced by others in the form of a hallucination. Proposed by British archaeologist Thomas Charles Lethbridge just over twenty years later, stone tape theory speculates that ghosts and hauntings are in fact mental impressions that have been released by living beings under extreme or traumatic circumstances and subsequently recorded by inanimate materials such as stone. If walls once had ears, they now also have mouths.

Given that the recordings are considered to be neither spectral nor otherworldly in nature, the theory implies that, under the right conditions, the recordings can be replayed and listened to. In this sense, ghosts are understood not as spirits but as noninteractive recordings, similar to the registration capacities of an audiotape machine that can play back previously recorded events. Of particular interest is the proposition that memories can be transferred from the somatic archive of the mind to the material ledgers of built and naturally occurring environments such as buildings, bridges and caves. In this way, memories—especially ones that are amplified via adrenaline, anxiety and fear—gain the capacity to change registers, from short-term to deep-time repositories. Or, as Dave Tompkins puts it, 'Tragedies of the past had been reduced to information in a continuous loop, putting the worst moment of your life on repeat' (Tompkins 2019).

On the Whispering Windows of the Peter Jones department store, the continuous loops on play bear no resemblance to the drama of the worst moments of one's life. Unless, of course, one considers the voyeuristic strains of capitalism, with its obsession for consumption, collection and curation, as mundane euphemisms for serial trauma. The pedestrian torrent of scopic desire registered by the windows is transferred and absorbed by the glass. Release only comes when its resonant frequency is transmitted. In an updated version of the stone tape, the flat-panel sound system is repurposed as a playback mechanism that revivifies the occulted capitalist archive stored in the window.

WINDOWS ON THE UNDEAD WEB

If the Whispering Windows are parallel vaults that register and display humanity's ocular compulsion, they could also be understood to be vibratory versions of the Web browser, since the endlessness of both negates depth and immersion. The browser window is a fluid interface of latitudes that refuses all borders. The computer's speakers connect with the Feonic technology, given that they both serenade the lateral skimming culture of hyperlink-hyperdesire culture. With their compressed atmospherics aquaplaning the high-end frequency range, they are perfect sound systems for the twenty-first century, acoustically reflecting the seductive flush-mount misery of ubiquitous promotion and consumption.

As part of almost everything on the earth's crust (90 per cent of it, to be exact), silicate minerals form the resonant stage upon which nature performs. After oxygen, silicon is the most abundant element. It is consistent, then, to find this ubiquity reflected in the built environment. Windows, building materials and computers are dependent on their silicon content. Given its significance in the composition of both material and digital realms, it could be posited that silicon is a conductor of human ambition and, dare we say, progress. This is an agency that orchestrates patterns of instrumentality. As the vibrating grains of cymatic experiments reveal structural drifts of frequencies, they also point to a collective intelligence—an Occidental sentience that is the equal of Negarestani's Middle Eastern oil in terms of scope, storage and conflict. The stone tapes—those static ledgers of distress—have been materially transformed, their chemical bonds shaken loose by the molten alchemy of capitalism and reconfigured in the form of glass. The near future will be mobilised by the shifting sands of Silica Tape theory.

Central to this new annealed premise is the fact that everything oscillates and has a resonant frequency. In this sense, it is simply a case of identifying the frequency in order to release the static and locked nature of an object. Once they are moving, objects start to tell different stories and to assume new relations with the world around them. Referencing David Bissell, Shelley Trower highlights his similar suggestion, that 'vibrations are *becomings* that undermine stable forms and identities' (Trower 2012, 8; emphasis original). Trower next goes on to note that, for Bruno Latour, 'vibration provides a model for "entities with uncertain boundaries, entities that hesitate, quake and induce perplexity"' (ibid.). This notion of uncertain boundaries is most pertinent when we consider the potential of any object to become a speaker, a transmitter of both internal (itself as a storage device) and external information. Vibration and oscillation become the keys to opening up the locked history of the archive and future of objects, along with being able to put objects into direct correlation with each other.

Expanding this notion, Trower goes on to propose that 'the idea of copresent frequencies puts time in space, in that the afterlife could be considered an area of the vibratory spectrum, the other world radiating through out world' (ibid., 49). Such a hypothesis is intriguing inasmuch as it opens up questions concerning the resonant frequencies of the future. This in turn implies that our ideas about linear time could also be unlocked by the key of resonance. In the introductory notes to their anthology *Unsound: Undead*, sonic-research unit AUDINT up the ante by suggesting that this bleeding of realms has been technologically instituted since the early twentieth century. They write, 'Ever since recording and communications technologies such as the phonograph and telephone have existed, there has been an interest in the potential of sound, infrasound and ultrasound to create domains of the undead, anomalous zones of transmission between realms of the living and the dead' (AUDINT 2019).

Of particular interest here are the ways in which waveforms are utilised to open channels to notions of otherworldliness. Duppies, electronic voice phenomena (EVP), voodoo, Santeria, candomblé, glossolalia, drones, static, white noise, chant, prayer, incantation, mantra—all of these phenomena are practiced on a global scale to achieve the blurring of this distinction between life/death, past/present/future and noise/silence. The hushed utterance of the Whispering and Talking Windows traverses all of these notions, untethering their meaning in the process. Vibration finds its voice of becoming in the form of the whisper. In hushed tones it articulates hidden agendas that set new dynamics into play. That which has been forged into solidity becomes fluid again. Excessive communication feeds back, muddying the clean signal, and Maurice Merleau-Ponty's lived body[1] becomes uninhabitable as it reneges on the real and its claims to veracity. Here the glacial emission of the undead—zombie sound—comes to represent not the guttural roar of anguish at being out of time but the quieted sibilance of time being shaken out of all that records it.

Let us loop back this silica tape theory so that we can further probe the undead applications of waveformed overload. Such an endeavour allows us to hear background sounds being woven into uprising—susurrations urging the suppressed back into the channels of (muted) public sound systems. While encouraging us to find new ways in which to parse the shell of the knowable—and thus to accommodate our modes of comprehension of that which is difficult to perceive—it also allows us to ask questions, the most pertinent of which regards what happens to the terminal public unsound system—the realm of the dead, which, in Friedrich Kittler's spectral equation, 'is as extensive as the storage and transmission capabilities of a given culture' (1986, 13), when it gets backed up.

In trying to answer this conundrum, we could do worse than consider the overflowing toilet in Francis Ford Coppola's 1974 film *The Conversation*. In the climactic scene of the movie, a pristine hotel room—which has possibly been the scene of a murder—is searched by a surveillance expert who is not sure as to whether this brutal outcome is what his recordings have captured. After finding that nothing is out of place in the room, he moves on to testing surfaces for any trace of wrongdoing, even the underneath of the bathroom plug, which proves to be clean. It is the diegetic sound of running water that finally leads the protagonist to engage in a last-ditch effort to flush the toilet in order to reveal the architecture's recent past—to expose what is hidden beneath its surfaces in the depths of excess management.

The subsequent torrent of blood that flows back up over the seat and onto the floor suggests that the life force of the departed woman is residing in the location where its death mask—faeces—is usually found. Mixing blood and shit together brings life and death into a material discourse, or, maybe better still, a dissolve that eliminates the boundary between the living and the dead. Returning to the question predicated on Kittler's thoughts about storage capacities, the above scene is the most base sonic and graphic version of what happens when they become full or their channels blocked. This is the oscillatory potential of everything in constant motion.

Blood flowing out of toilets, taps and baths has become a sonic and visual trope many film directors turn to. Most regularly it is deployed to suggest that things are not only out of kilter but, more than that, that the mechanisms in place to separate the excess from the functional, the dead from the living, have been taken over by some imperceptible agency. The natural movement of things—garbage, shit, bodies—is reversed so that they flow back into environments only worthy for those considered to be sanitised, civilised, living. And those who encounter such occurrences become privy to the existence of this otherworld, the agencies manifest within it and the realisation that they will have to successfully navigate the boundary between worlds in order to survive.

As discussed throughout this book, the waveformed domain is regularly cited as a domain in which portals to other worlds are opened up. Transmission technologies, from wired to bone-conduction speakers, form a distributed interface to such channels. Such fluid mechanisms also allow us to start decrypting the encoded future trajectories of vibration, technology and sentience—reflecting both promising and insidious potentials in the process. As an oscillating apparatus that portends future formats of communication and cognition, the Whispering and Talking Windows signal how the antenna-body will develop.

If in this context of the otherworldly the zombie is the somatic manifestation of excessive signals—the fleshed-out feedback from existence—then the whisper is its disembodied call from beyond. And it is from beyond that future frequency-based agency will feedback into everyday living, all sound and unsound speakers linked through space and time in a massive distributed system—an oscillating intelligence that will come to redefine the breadth of waveformed pressure along with our comprehension of the ways in which speaker systems are used to influence, manipulate and torture.

NOTE

1. French phenomenologist Maurice Merleau-Ponty conceived of this subject body in order to express ideas about the role that perception plays in interpreting being-in-the-world. The *lived body* refers to how the nature of the body affects perception in a 'real' world of intricately entwined realities (Merleau-Ponty 1962).

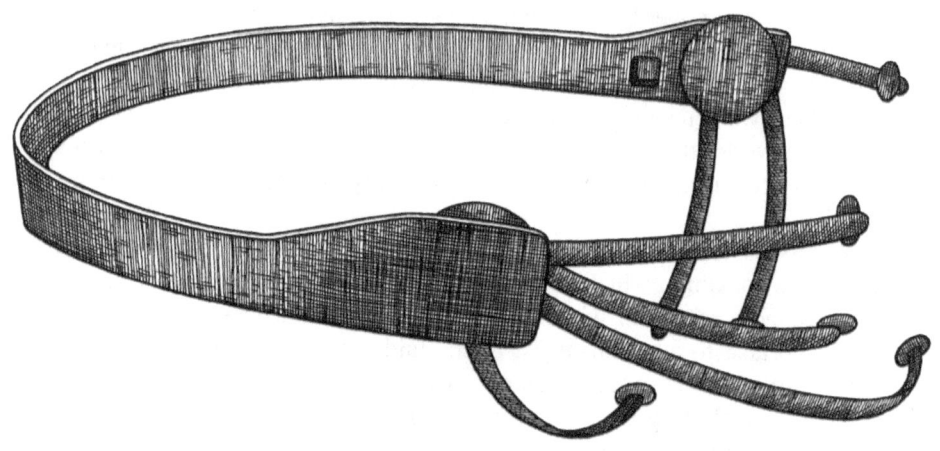

MIT Media Lab–AlterEgo.
Illustration by Krystian Griffiths

Conclusion
Phantom Sound Systems

The conclusion to *Sound Pressure* does not only work in the traditional manner of neatly summing up or tying loose ends together, it also gains purchase as an ending by concentrating on the beginnings of two new sound systems in their infancy at the time of this writing—the unknown technology responsible for the alleged acoustic attacks in Cuba and China and the AlterEgo device being developed at MIT. While the capacity of neither of these technologies is currently fully understood, what is known is that they both whisper details of the future trajectories of stealth sonics. They do this by not just extending our current notion of the sensorium but by also channelling waveformed imminence—as an augury of the power of sonic fiction, especially when it is weaponised and perfectly composed to interface with the new forms of sentience emerging from artificial and diminished intelligences.

ACOUSTIC ATTACK: CUBA

In terms of abstracting and exploding our traditional notions of what a sound system is and how it functions, there are few examples that extend our expectations beyond the dynamics of disappearance performed by Whispering Windows—that is, unless the act of materialisation and dematerialisation is played out in language only, which brings us to a waveformed technology that exists through its capacity to function in the collective anxiety and consciousness of a population. This transmission technology is possibly the last word in mythical sound systems and the first words of the twenty-first century that truly bespeak the sociopolitical resonance of sonic fiction on a world stage. Replacing the component parts of drivers, boxes and amplifiers with hyperstition, conspiracy theory and speculation, this is the C21st Unsound System par excellence. This is the phantom speaker system that allegedly caused the Cuban silent acoustic attacks on the US embassy in Havana in 2017.

'On Aug. 23, CBS News reported having obtained and reviewed medical records for the US diplomats that showed they had experienced such symptoms as hearing loss, nausea, headaches, and balance problems and were diagnosed with conditions as serious as mild traumatic brain injury' (Rubin 2017). The next day, on the 24th, Heather Nauert, spokesperson for the US State Department, alerted reporters to the news that at least sixteen US government employees (which would later rise to twenty-two in number confirmed affected) along with a Canadian diplomat had been affected, with the 'attacks' having first started back in 2016.

The Cuban government vociferously denied any involvement. Foreign Minister Bruno Rodríguez Parrilla went further, claiming that the US government was being purposefully duplicitous and that the incident was being used as a 'political pretext for damaging bilateral relations and eliminating the progress made' when the Obama administration had been in power. This was not business as usual, even for countries whose often-outlandish narratives of anxiety about each other have skittered along delicate lines of communications and conflict.

The reportage continued, 'Speculation grows that the diplomats were covertly targeted with some new device that uses sound as a weapon' (Rubin 2017). And so started a global memeplex that connected village rags and blogs to the largest media conglomerates. Sonic weapons were the first culprits but were quickly displaced by ultrasonic or infrasonic technologies. The memeplex's limbs of propagation grew multifold, akin to an infomatic version of *Tetsuo, the Iron Man* (Tsukamoto 1989). As more news channels around the world began transmitting the story, their poor fact-checking caused the story to mutate and fuse with the environmental disinformation that would become part of the narrative. The by-now phantom Unsound System was becoming Frankensteinian in its assembly. A quotation here, a factoid there, all glommed together by the sweat of a culturally nervous disposition.

Researchers in the field, such as aforementioned Jürgen Altmann, expressed their doubts about the existence of such a device. 'According to my research, strong effects on humans require loudness levels that would be perceived as very loud noise while exposed. . . . And projecting ultrasound over long distances and or through walls or closed windows is difficult' (Rubin 2017). Other experts and analysts in the field, such as Joseph Pompei of Holosonic Research Labs, Inc., and myself, were regularly contacted and interviewed for opinion and further speculation about the types of frequencies and technologies that could be responsible. Add the hypothesis that Russia could be behind the presumed hostility (whom the US government accused in a prime instance of reverse-maskirovka), and it is easy to understand how the hot-off-the-press monster was brought to life: the changes in air pressure caused by the phantom Unsound System chilling the atmosphere of a new cold war. These are the sounds of the dormant enemy being resurrected.

After months of being informed by numerous researchers that acoustic or nonacoustic weapons were probably not responsible, reporters looked for new explanations. The Trump administration was unwavering, however, in its belief that Cuba was responsible, Trump stating as much in a news conference with Senate Majority Leader Mitch McConnell in the White House Rose Garden on 16 October. Only days before this, on the 12th, *The Guardian* ran the article 'Mass Hysteria May Explain "Sonic Attacks" in Cuba, Say Top Neurologists' (Borger and Jaekl 2017), and the story shifted emphasis—at least in terms of the blame game.

According to this article and a slew that followed over the ensuing months, the diplomatic rift between Cuba and the United States had been caused neither by intentional attack nor by faulty surveillance equipment accidentally emitting high frequencies, as had also been suggested. '"US and Cuban investigations have produced no evidence of any weapon. . . . From an objective point of view, it's more like mass hysteria than anything else", said Mark Hallett, the head of the Human Motor Control Section of the US National Institute of Neurological Disorders and Stroke' (ibid.). Slowly over the following weeks and months, into early 2018, the conjecture wove itself around this newly posited logic of psychosomatic disease before dissipating into an ambient infection that sat in the sore of US–Cuban relations.

ACOUSTIC ATTACK: CHINA

To all intents and purposes, the story revolving around the Cuban Unsound System had been abstracted, diffused and played out. Then, on 23 May 2018, it turned out that the story had not met its demise and been left to rot in the 'bizarre and unanswered' category. It had in fact merely changed location—to China. The phantom Unsound System had been revivified to play one last time. According to the US State Department, a staff member in the American consulate in Guangzhou complained of experiencing 'abnormal sensations of sounds and pressure' (Sky News 2018), which led to officials declaring the staffer had suffered mild traumatic brain injury. Stating concern over this 'serious health incident', US Secretary of State Mike Pompeo redeployed the frequency-based-weapon hypothesis. Armed and loaded, the notion of unsound had been weaponised again in the bellicose auditorium of US foreign policy.

The response of the US State Department was written in typical Trumpian prose, the final sentence of advice being especially symbolic of its strategy to strip out any content that could be useful and blunting communication to the point of being aberrant. The health alert issued to US staff reads: 'While in China, if you experience any unusual acute auditory or sensory phenomena accompanied by unusual sounds or piercing noises, do not attempt to locate their source. Instead, move to a location where the sounds are not present' (Sky News 2018). What is continually present here is the political and economic contexts in which these events have occurred. With the United States and China on the cusp of a trade war over increased tariffs, whether planned or not, the anxiety over the economic downturn is partially channelled by unsound narratives that clearly identify and demarcate the enemy's future subterfuge.

Presently the Chinese waveformed event has only just started rolling out its waves of paranoia. It is safe to say that, since the initial State Department press statement, next to nothing has been learned or at least publicised about the frequency-based technology that might or might not be responsible for these liminally perceptive hostilities. What we do know is that the predictions made about contemporary and future conflict being more asymmetric and pervasive in quotidian circumstances are proving to be accurate. The Cuba/Chinese Unsound System—or CCUS—updates the delivery method of the sound object[1] as a waveformed phenomenon free from its origin and, along with other technologies examined in this book, heralds the sensorium as an emergent location for conflict.

As a transmission mechanism, the CCUS exists at the intersection of anxiety, fantasy and conspiracy theory. To comprehend it fully requires a versatile and responsive mode of enquiry—a conceptual technique not that different from Henri Lefebvre's designated brand of theory that 'render[s] intelligible qualities of space which are at once perceptible and imperceptible to the senses' (Merrifield 2000, 170). In this way the potentiality of the unsound system to emit frequencies that move beyond the perceivable has itself conducted a manoeuvre that alchemically transforms its own material being from matter to (mis)information. It carries the threat of affecting individuals and groups of people anywhere at anytime, even in highly protected and surveilled buildings, such as embassies. Where flat-panel technologies relocate speaker resonance to surfaces, the CCUS operates most effectively behind the scenes in the darkened theatres of realpolitik.

SUBVOCALISATIONS OF THE ALTEREGO

If the Cuba/Chinese Unsound System is the ultimate manifestation of unsound arrangements on a world stage, then the AlterEgo device—currently under development at the Massachusetts Institute of Technology—is its internalised twin. So, whereas the CCUS brings the liminal to bear on the body from and through distance, the AlterEgo delivers intramural resonance out of its somatic casing and into discreet and seamless conversation with a machine, for this resonant mechanism allows a user to access and amplify that most-covert and -camouflaged of articulations—the inner voice of the other.

MIT's overview webpage for the AlterEgo offers further clarification as to its purpose: It is 'a closed-loop, noninvasive, wearable system that allows humans to converse in high-bandwidth natural language with machines, artificial intelligence assistants, services, and other people without any voice—without opening their mouth and without externally observable movements—simply by vocalising internally. The wearable captures electrical signals, induced by subtle but deliberate movements of deliberate internal speech articulators (when a user intentionally vocalises internally), in likeness to speaking to one's self' (Massachusetts Institute of Technology 2018b).

Explaining how the AlterEgo acts as a seamless interface for the data/body, the text continues, 'We use this to facilitate a bidirectional natural-language computing system, where users receive aural output through bone-conduction headphones without obstructing a user's physical senses. This enables a user to transmit and receive streams of information to and from a computing device or any other person without any observable action, in discretion and without invasion of the user's privacy. AlterEgo aims to combine humans and computers—such that computing, the Internet, and AI would weave into human personality as a "second self" and augment human cognition and abilities' (ibid.).

The process of saying words in your head when writing or reading, for instance, is known as *subvocalisation*, *silent speech* or *auditory reassurance*. This type of internal articulation produces minute gestures in the larynx and in other muscles associated with speaking. It is these movements, known as *neuromuscular* or *endogenous electrical signals*, that are picked up by the AlterEgo's seven surface-based electrodes that line either side of the mouth and the jaw. These granular gestures are processed by a modular neural network-based pipeline trained to detect and recognise the muted expressions, which are then reconstructed as words—a preemission audile technique[2] for the twenty-first century. This is the emergence of silent speech as machine interface, subvocalisation as a sounding board of the auditive threshold—an intimate transmutation of speech recognition.

As such, the AlterEgo—along with other recent systems like those being developed by neuroengineers at Columbia University that harness AI to translate signals from the brain into audible speech (Akbari et al. 2018)—might be more accurately defined as *sounding systems*, extending the operations of a recording device or a speaker, dislocating our preconceptions about what constitutes transmission. Representing another shift in the mutating identity of the antenna-body, it still retains its main characteristics, transmitting and receiving silently, problematising any fixity of perception, channelling the inside out and the outside in. Plugging bio and digital networks into each other's matrices of extension, the internal articulation and the AI assistant's voice engage in unspoken relations for the first time.

One of the most pressing questions to arise when considering emergent modes of behaviour afforded by the AlterEgo concerns its future updates. It is already imagined that prospective versions could inaudibly consult with 'clinical decision-making AI agents' to ameliorate medical servicing. How long will it be, one might ask, before the internal voice can take the place of the AI voice, and, conversely, how long will it be before the voice of an AI can interfere with, or supplant, the host channel of intimate articulation? Michel Serres' (1982) invocation of the fluid lines between parasite and host again comes to mind here, as the circumstances for them to change positions become increasingly personalised and relevant outside of the computer/body dynamic.

Establishing communication at the intersection of bone conduction, audio intelligence and extended sensory augmentation, the AlterEgo is a waveformed exemplar of technological convergence. '"The motivation for this was to build an IA device—an intelligence-augmentation device", says Arnav Kapur, a graduate student at the MIT Media Lab who led the development of the new system. "Our idea was, Could we have a computing platform that's more internal, that melds human and machine in some ways and that feels like an internal extension of our own cognition?"' (Hardesty 2018).

Along with the amplification of cognition, the AlterEgo technology and, more specifically, its bone-conduction headphones enable and encourage the recalibration of the sensorium. Chiming with Fredric Jameson's exhortation 'to grow new organs, to expand our sensorium and our body' (1991, 80), such augmented/audio intelligence presents us with another clue as to how future modes of communication will bypass the traditional employment of sensory mechanisms. It is a suggestion that ameliorates Maurice Merleau-Ponty's model of the habitual body[3] from being impelled to act in a recurrent loop of cause and effect and instead urges for a new generation of informational dynamics to renegotiate comprehensions of being in the world. The fewer somatic editing suites that sit between stimuli and neurological processing, the more seamless and less noisy the interaction—or, at least, that is how such sensorial systems are being positioned for market.

On the other end of the public-relations spectrum, the following proclamation offers further pointers as to how the technology will be eased into public consciousness and uptake: 'It does not read the thoughts coming up in the user's mind, only the ones a user consciously intends to send to the device. . . . It is crucial that the control over input resides absolutely with the user in all situations', reads one of the answers on the FAQ page of the device's MIT-based website (Massachusetts Institute of Technology 2018a). Guessing—probably accurately—that readers might be vaguely itching with a *Scanners*-induced anxiety (Cronenberg 1981), their following statement, while meaning to allay such fears, only brings to mind more forcefully the technology-run-amok sci-topia of 1970s films: 'No, this device cannot read your mind' (Massachusetts Institute of Technology 2018a).

But this depends on how your regard the seat of consciousness, especially if you believe in distributed cognition and that 'the mind, beyond subjective experience and beyond conscious and nonconscious information processing, can be seen as a self-organizing, emergent process of a complex system. And that system is both within us and between us and others' (Siegel 2014). Rather than concentrating on exacting definitions of *consciousness*, the most telling aspect of this statement is the way it induces questions about the functionality of future upgrades and the types of sectors that the sounding system will service.

This is a long way from the logic of Muzak in the factory where work rhythms were broken down into atomised gestures or during the Waco siege where relations between members of the Branch Davidian were studied and split for tactical gain. The AlterEgo is, however, concerned with gestures of an intensely discreet nature, and when predictive analytics and pressure are thrown into the equation, one does not have to engineer too radical a narrative to imagine that certain sections of the military and entertainment industries might find ulterior uses for a technology that can access language and ideas without having to utilise the outer ear, eyes or mouth.

PRESSURE AS MIMETIC ORDNANCE

Both of the speculative systems that have been used as denouements to this study of pressure-based technologies have been employed as harbingers of divergent yet connected unsound futures—projected domains in which the old types of spaces once considered protected from waveformed agency are laid vulnerable to the vagaries of frequencies. While one system signals an external pressure based on phantom relations, delivering the transgressive capacity of unsound into the political safe haven of the embassy, the other represents the evasive and obliquely defended zone of the inner voice. No one and nowhere is safe from harm. They probably never were.

Increasing the pertinence of this statement are the implications regarding inner sanctuaries, for it is these spheres that are being transgressed and put under pressure by unsounding systems. The inner voice and the embassy building find some topological equivalence here, at the margins of that which is off limits. These are spheres that external presences and pressures are not supposed to infiltrate, domains where covert aspirations run deep and shape language in the name of the untouchable. Through this prism the inner voice represents the embassy of the unspoken. And, in response, the embassy becomes the inner voice of foreign relations—residing and hiding within its host country, in plain sight, hearing and touch.

Frequency-based technologies, especially more recently emerging ones that have more complicated relationships with perception, trade in this kind of breached currency—a type of waveformed exchange that demands an open and converged flow of information, ethics and relations. Losing their sense of solid ground, these latter markers of access, consciousness and space become liquid assets in the most chaotic sense. *Liquid agency* might be a better term, in fact, for the nature of waveformed systems that bring pressure to bear on anything that might be considered untouchable, fixed and isolated. This is why such oscillating technologies are utilised. Vibration has no time for boundaries. It places everything into a connective skein and plunges old alliances into a dissolving economy of association, with materiality and non-materiality becoming merely different channels of propagation.

From a more expansive perspective, how connection—its corollary effects and affect—and isolation are woven into sound, unsound and unsounding-systems functionality is central to the way that technologies and events have been analysed throughout this text. Thus, as much as this book's directive has been to understand how such systems isolate, it also inevitably raises questions around how they connect beyond the obvious pleasure/joy/empathy axis. An alternate approach to this enquiry would be to ask what happens when we are overconnected, when we have too many affinities within the social, technological and vibratory mesh that envelops us.

On a daily basis we are connected to a plethora of networks—AI assistants, distributed intelligences, augmented/diminished/virtual realities, Internets of things and multitudes of databases that document our interactions with banks, travel agents and government agencies. As much as connection offers us in terms of vibratory information, it also serves to isolate us in more extreme ways than we have ever known. The drive to separate and alienate runs in direct correlation to the drive to connect, and much of what seems to be in question right now with regards to the vibrational/digital mesh is based on how we position the notion of the individual in relation to the collective.

The capacity to weaponise modes of isolation and connection increases the more extreme either situation becomes. The mimetic ordnance of the CCUS waveformed attacks and extrusion of the inner voice point to both ends of the spectrum of oscillation—of being separated and intrinsically singular as one deals inwardly with trauma on the one hand and appended to the binary tissue of the Internet through the granularity of micro gestures on the other. It is this involute tension between isolation and connection that ultimately resides at the heart of the sound/unsound/unsounding system—the operational modulation between relational states the only intrinsic metric of value.

A good example of this dynamic is the sound system used within the blacked-out Guantánamo cells. It symbolises a dark inversion of those technologies used by musical subcultures, its purpose not only to break the identity of its captive listeners but also to go further and orchestrate their cognitive malfunction. From a sociopolitical standpoint, the sound system 'comes before the subject with heightened intensity, bearing a mysterious charge of affect, here described in the negative terms of anxiety and loss of reality, but which one could just as well imagine in the positive terms of euphoria, the high, the intoxicatory or hallucinogenic intensity' (Jameson 1991, 73).

It is the sensorial arrangement of the sound system—on the periphery of the vacillating sublime, about to descend its listeners (at the flick of a switch) into realms of decadent pleasure or into the abyss of hellish torture—that reveals its bipolar potential. Being able to dial into this channel has always been on the agenda of those who have wanted to break free of the material and spatial constraints of surround-sound techniques. The creations of the HSS and the Whispering and Talking Windows are the realisation of such aspirations for they are technologies that can elide architectonic parallels and allow military, policing and entertainment-based organisations to directly do what they have always wanted to do—namely, get into another's head.

No other phenomena render the conditions of fluid agency more apparent than waveforms. As such, the future of vibration will be mined more deeply and asymmetrically for its facility to be channelled by technologies that can activate and extend both potentials—impending oscillation at a perceived and noncognisant level being a volatile agent that breaks down the distinction between the inner and externalised voice, press statement and meme, rest and sleep. And the pressure necessary to enact such a deliquescent sensibility comes easily to the antenna-body. Pressure as a flow of the static. Pressure that reorients a cartography that is 'organic and fluid and alive; it has a pulse, it palpitates, it flows and collides with other spaces' (Merrifield 2000, 170). Pressure that is sound, unsound and unsounding all at once.

NOTES

1. Pierre Schaeffer first coined the term *sound object* in 1966. Becoming an influential idea within music academia, a sound object was later interpreted by Michel Chion (1983) to mean any sonic happening that could be interpreted as a coherent sum, irrelevant of the production of the sound or the interpretation of it.

2. Jonathan Sterne conceives of this term in his book *The Audible Past: Cultural Origins of Sound Reproduction* (2002). *Audile technique* refers to a skill for distinguishing the nature of sounds, a capacity historically embodied in the work of two exemplary professions—the physician and the telegrapher.

3. Maurice Merleau-Ponty conceived of this subject body in order to express ideas about the role perception plays in interpreting being-in-the-world. The *habitual body* is a somatic form that receives information and acts upon it according to previous memories of same or similar activities, in relation to external stimuli. In this way, previous modes of experience sediment themselves into the behaviouralisms of the habitual body.

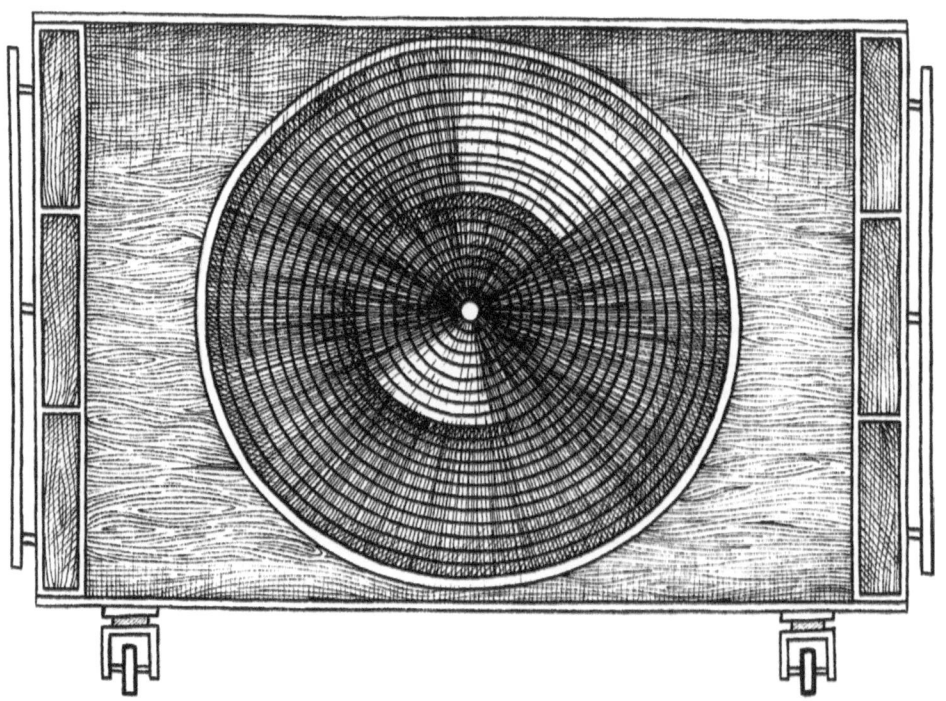

The Audaunter–Powersoft M-Force Subwoofer.
Illustration by Krystian Griffiths

References

WORKS CITED

Abe, D., M. Arai, and M. Itokawa. 2017. 'Music-Evoked Emotions in Schizophrenia'. *Schizophrenia Research* 185: 144–47.

AC/DC. 1979. *Highway to Hell*. Atlantic Records. LP Record.

Adorno, Theodor. 1973. *Philosophy of Modern Music*. Translated by Anne G. Mitchell and Wesley V. Blomster. New York: The Seabury Press.

Akbari, Hassan, Bahar Khalighinejad, Jose Herrero, Ashesh Mehta, and Nima Mesgarani. 2018. 'Reconstructing Intelligible Speech from the Human Auditory Cortex'. *bioRxiv*, Cold Spring Harbor Laboratory, 19 June. https://www.biorxiv.org/content/biorxiv/early/2018/10/10/350124.full.pdf.

———. 2019. 'Towards Reconstructing Intelligible Speech from the Human Auditory Cortex'. *Nature*, 29 January. https://www.nature.com/articles/s41598-018-37359-z.

Allen, John. 2000. 'On Georg Simmel: Proximity, Distance and Movement'. In *Thinking Space*, edited by Mike Crang and Nigel J. Thrift, 54–70. London: Routledge.

Altmann, Jürgen. 1998. 'Acoustic Weapons: A Prospective Assessment'. *Science & Global Security* 9, no. 3: 165–234.

———. 1999. 'Acoustic Weapons? Sources, Propagation, and Effects of Strong Sound'. ASA/EAA/DAGA Meeting, Berlin, 17 March. acoustics.org/pressroom/httpdocs/137th/altmann.html.

Amnesty International. 2007. 'United States of America: Cruel and Inhuman; Conditions of Isolation for Detainees at Guantánamo Bay'. April. https://www.amnesty.ie/wp-content/uploads/2016/04/Guantanamo-Conditions.pdf.

Associated Press. 2003. 'Ames Company Develops Device That Emits Sound from Windows, Walls'. *Globe Gazette*, 2 August. http://globegazette.com/business/ames-company-develops-device-that-emits-sound-from-windows-walls/article_42a56b21-6e1e-5940-ba03-83a1ffba0081.html.

Attali, Jacques. 1985. *Noise: The Political Economy of Music*. Translated by Brian Massumi. Manchester: Manchester University Press.
AUDINT [Steve Goodman, Toby Heys, and Eleni Ikoniadou], eds. 2019. *Unsound: Undead*. Falmouth: Urbanomic x Sequence Press.
Augé, Marc. 1995. *Non-places: Introduction to an Anthropology of Supermodernity*. Translated by John Howe. London: Verso.
AZoM. 2002. 'Transducer Materials for Sonar Systems—Materials Comparison'. AZO Materials (website). 30 April. https://www.azom.com/article.aspx?ArticleID=1377.
Babbage, Charles. 1835. *On the Economy of Machinery and Manufacturers*. London: Charles Knight.
Ballard, J. G. 1960. 'The Sound-Sweep'. *Science Fantasy* 13, no. 39.
Bassett, Caroline. 2003. 'How Many Movements?' In *The Auditory Culture Reader*, edited by Michael Bull and Les Back, 343–56. Oxford: Berg.
Bateson, Gregory. 1973. *Steps to an Ecology of Mind*. London: Granada.
Baudrillard, Jean. 1991. *La Guerre du Golfe n'a pas eu lieu*. Paris: Galilée.
Bayoumi, Moustafa. 2005. 'Disco Inferno'. *The Nation*, 8 December. http://www.thenation.com/article/disco-inferno.
BBC News. 2003a. 'Sesame Street Breaks Iraqi POWs'. Last updated 20 May. http://news.bbc.co.uk/2/hi/middle_east/3042907.stm.
———. 2003b. 'The Weapons of Bloodless War'. Last updated 13 May. http://news.bbc.co.uk/go/pr/fr/-/2/hi/technology/3021873.stm.
———. 2006. 'Triple Suicide at Guantánamo Camp'. 11 June. http://news.bbc.co.uk/2/hi/americas/5068228.stm.
Beatles, The. 1968. *The Beatles* [*The White Album*]. Apple Records. LP Record.
Beaupeurt, J. E., F. W. Snyder, S. H. Brumaghim, and R. K. Knapp. 1969. *Ten Years of Human Vibration Research: Human Factors Technical Report*. August, Office of Naval Research, Wichita, KS: The Boeing Company. https://apps.dtic.mil/dtic/tr/fulltext/u2/693199.pdf.
Bhabha, Homi K. 1994. *The Location of Culture*. London: Routledge.
Bijsterveld, Karin. 2003. 'The Diabolical Symphony of the Mechanical Age: Technology and Symbolism of Sound in European and North American Noise Abatement Campaigns, 1900–40'. In *The Auditory Culture Reader*, edited by Michael Bull and Les Back, 165–89. Oxford: Berg.
Bishop, Jack. 2002. 'Schismogenesis? The Global Industrialization of Brazilian Popular Music'. *Associação Brasileira de Etnomusicologia (ABET)*, Recife, Pernambuco, Brasil, 20 November.
Borger, J., and P. Jaekl. 2017 'Mass Hysteria May Explain "Sonic Attacks" in Cuba, Say Top Neurologists'. *The Guardian*, 12 October. https://www.theguardian.com/world/2017/oct/12/cuba-mass-hysteria-sonic-attacks-neurologists.
Bromley, David G., and Edward D. Silver. 1995. 'The Davidian Tradition: From Paternal Clan to Prophetic Movement'. In *Armageddon in Waco: Critical Perspectives on the Branch Davidian Conflict*, edited by Stuart A. Wright, 43–72. Chicago: University of Chicago Press.
Brown, Mick. 2007. *Tearing Down the Wall of Sound: The Rise and Fall of Phil Spector*. London: Bloomsbury.

Bull, Michael, and Les Back. 2003. 'Introduction: Into Sound'. In *The Auditory Culture Reader*, edited by Michael Bull and Les Back, 1–25. Oxford: Berg.
Burgess, Anothony. 1962. *A Clockwork Orange*. London: Heinemann.
Cahn, Edward L., dir. 1959. *Invisible Invaders*. USA: United Artists.
Cannon, Walter B. 1915. *Bodily Changes in Pain, Hunger, Fear and Rage: An Account of Recent Researches into the Function of Emotional Excitement*. New York: D. Appleton and Co.
Chion, Michel. 1983. *Guide des objets sonores: Pierre Schaeffer et la reserche musicale*. Paris: INA-GRM/Buchet-Chastel.
Cixous, Helène. 1970. *The Third Body*. Translated by Keith Cohen. Evanston, IL: Northwestern University Press.
Coetzee, J. M. 1988. *White Writing: On the Culture of Letters in South Africa*. New Haven, CT: Yale University Press.
Cole, Ronald H. 1995. *Operation Just Cause: The Planning and Execution of Joint Operations in Panama, February 1988–January 1990*. Washington, DC: Joint History Office, Office of the Chairman of the Joint Chiefs of Staff. https://www.jcs.mil/Portals/36/Documents/History/Monographs/Just_Cause.pdf.
Connor, Steven. 2004. 'Topologies: Michel Serres and the Shapes of Thought'. *Anglistik* 15: 105–17.
Conroy, John. 2000. *Unspeakable Acts, Ordinary People: The Dynamics of Torture*. Berkeley: University of California Press.
Coppola, Francis Ford, dir. 1974. *The Conversation*. USA: Paramount Pictures.
Corbin, Alain. 1998. *Village Bells*. Translated by Martin Thom. New York: Columbia University Press.
Crang, Mike, and Nigel Thrift. 2000. 'Introduction'. In *Thinking Space*, edited by Mike Crang and Nigel Thrift, 1–30. London: Routledge.
Crary, Jonathan. 1992. *Techniques of the Observer: On Vision and Modernity in the Nineteenth Century*. Cambridge, MA: MIT Press.
Crimzon Zone. *Iraqi Gothic*. Self-released. Compact disc.
Crocker, Stephen. 2007. 'Noises and Exceptions: Pure Mediality in Serres and Agamben'. *CTHEORY*. http://www.ctheory.net/articles.aspx?id=574.
Cronenberg, David, dir. 1981. *Scanners*. Canada: AVCO Embassy Pictures.
———, dir. 1983. *Videodrome*. Canada: Universal Pictures.
Cusick, Suzanne G. 2006. 'Music as Torture/Music as Weapon'. *Revista Transcultural de Música/Transcultural Music Review* 10. https://www.sibetrans.com/trans/articulo/152/music-as-torture-music-as-weapon.
Debord, Guy-Ernest. 1955. 'Introduction to a Critique of Urban Geography'. Translated by Ken Knabb. *Les Lèvres Nues* 6 (September). Retrieved from http://library.nothingness.org/articles/SI/en/display/2.
De Certeau, Michel. 1984. *The Practice of Everyday Life*. Translated by Steven Rendall. Berkeley: University of California Press.
DeLanda, Manuel. 1991. *War in the Age of Intelligent Machines*. New York: Zone.
Deleuze, Gilles, and Félix Guattari. 2004. *Anti-Oedipus: Capitalism and Schizophrenia*. Translated by Robert Hurley, Mark Seem, and Helen R. Lane. London: Continuum.

Dorey, Peter. 1995. *The Conservative Party and the Trade Unions*. London: Routledge.
Dowdle, John Erick, and Drew Dowdle, creators. 2018. *Waco*. Television. USA: Paramount Network.
Durrett, Deanne. 1998. *Unsung Heroes of World War II: The Story of the Navajo Code Talkers*. New York: Facts on File.
Dyer, Richard. 1979. 'In Defence of Disco'. *Gay Left* Summer, no. 8: 20–23. http://gayleft1970s.org/issues/gay.left_issue.08.pdf.
Ellison, Christopher G., and John P. Bartowski. 1995. 'Babies Were Being Beaten: Exploring Child Abuse Allegations at Ranch Apocalypse'. In *Armageddon in Waco: Critical Perspectives on the Branch Davidian Conflict*, translated by Stuart A. Wright, 111–49. Chicago: University of Chicago Press.
Erikson, Kai. 1966. *Wayward Puritans*. New York: Wiley.
Fackler, Guido. 2003. '"We All Feel This Music Is Infernal . . .": Music on Command in Auschwitz'. In *The Last Expression: Art and Auschwitz*, edited by David Mickenberg, Corinne Granof, and Peter Hayes, 114–25. Chicago: Mary and Leigh Block Museum of Art/Northwestern University.
Ford, Henry, and Samuel Crowther. 1922. *My Life and Work*. New York: Garden City Publishing Company.
Foucault, Michel. 1966. *Les mots et les choses: Une archéologie des sciences humaines*. Paris: Éditions Gallimard.
———. 1975. *Discipline and Punish: The Birth of the Prison*. Translated by Alan Sheridan. New York: Random House.
———. 1980. *Power/Knowledge: Selected Interviews and Other Writings 1972–1977*. Edited by Colin Gordon. London: Harvester Press/Pantheon Books.
———. 2002. 'Of Other Spaces: Utopias and Heterotopias'. In *The Visual Culture Reader*, 2nd ed., edited by Nicholas Mirzoeff, 228–36. London: Routledge.
4th25. 2005. *Live from Iraq*. Self-released. Compact disc.
Freeman, John. 2016. 'Tupac's "Holographic Resurrection": Corporate Takeover or Rage against the Machinic?' *CTHEORY*. http://ctheory.net/tupacs-holographic-resurrection-corporate-takeover-or-rage-against-the-machinic/.
Friedman, Herbert A. 2009. 'The "Wandering Soul" Tape of Vietnam'. *Psywarrior*. Accessed 8 January. http://pcf45.com/sealords/cuadai/wanderingsoul.html.
Frisby, David Patrick, and Mike Featherstone, eds. 1997. *Simmel on Culture: Selected Writings*. Thousand Oaks, CA: Sage.
Fuller, Matthew. 2005. *Media Ecologies: Materialist Energies in Art and Technoculture*. Cambridge, MA: MIT Press.
Gaffney, Edward McGlynn, Jr. 1995. 'The Waco Tragedy: Constitutional Concerns and Policy Perspectives'. In *Armageddon in Waco: Critical Perspectives on the Branch Davidian Conflict*, edited by Stuart A. Wright, 323–58. Chicago: University of Chicago Press.
Geneva Convention (III) Relative to the Treatment of Prisoners of War, 12 August 1949. 1949. 'Section I, Questions of Prisoners, Article 17'. International Committee of the Red Cross. https://ihl-databases.icrc.org/applic/ihl/ihl.nsf/Article.xsp?action=openDocument&documentId=2D8538058860C1FCC12563CD0051ABBE.

Gerard, Philip. 2002. *Secret Soldiers: The Story of World War II's Heroic Army of Deception*. New York: Dutton.

Gilmore, John. 1995. *Cold-Blooded: The Saga of Charles Schmid, the Notorious Pied Piper of Tucson*. Los Angeles: Feral House.

Goldberg, Ken, ed. 2000. *The Robot in the Garden: Telerobotics and Telepistemology in the Age of the Internet*. Cambridge, MA: MIT Press.

Goodman, Steve. 2009. *Sonic Warfare: Sound, Affect, and the Ecology of Fear*. Cambridge, MA: MIT Press.

———. 2014. 'Indiscernible Loudspeakers: When Things Talk'. In *Booster: Art Sound Machine*, edited by Marta Herford Museum for Art, Architecture, Design. Exhibition catalogue. Bielefeld, Germany: Kerber Verlag.

Gorriors of Ragnarok. 2007. *Rape Guts*. Splatter/Suck Records. Compact disc.

Gramsci, Antonio. 1999. *Selections from the Prison Notebooks of Antonio Gramsci*. Translated and edited by Quintin Hoare and Geoffrey Nowell Smith. New York: International Publishers.

Hall, John R. 1995. 'Public Narratives and the Apocalyptic Sect: From Jonestown to Mt. Carmel'. In *Armageddon in Waco: Critical Perspectives on the Branch Davidian Conflict*, edited by Stuart A. Wright, 205–35. Chicago: University of Chicago Press.

Hambling, David. 2008. 'Microwave Ray Gun Controls Crowds with Noise'. *New Scientist*, 3 July. http://www.newscientist.com/article/dn14250-microwave-ray-gun-controls-crowds-with-noise.html.

Hardesty, Larry. 2018. 'Computer System Transcribes Words Users "Speak Silently"'. *MIT News*, 4 April. http://news.mit.edu/2018/computer-system-transcribes-words-users-speak-silently-0404.

Harvey, David. 2001. *Spaces of Capital: Towards a Critical Geography*. New York: Routledge.

Hebdige, Dick. 1979. *Subculture: The Meaning of Style*. London: Routledge.

Henriques, Julian. 2003. 'Sonic Dominance and the Reggae Sound System Session'. In *The Auditory Culture Reader*, edited by Michael Bull and Les Back, 451–80. Oxford: Berg.

———. 2011. *Sonic Bodies: Reggae Sound Systems, Performance Techniques, and Ways of Knowing*. London: Bloomsbury.

Heys, Toby, and Andrew Hennlich. 2010. 'The Art of Conservative Détournement'. *ETC* 88: 61–64.

Hitchcock, Alfred, dir. 1940. *Foreign Correspondent*. USA: United Artists.

Holosonics. 2009. 'Audio Spotlight by Holosonics'. http://www.holosonics.com/. Accessed 14 February.

———. 2019a. 'Frequently Asked Questions'. Accessed 28 March. https://www.holosonics.com/faq.

———. 2019b. 'What Makes a Sound Source Directional?' Accessed 28 March. https://www.holosonics.com/what-makes-a-sound-source-directional.

Hultkrans, Andrew. 2008. 'The Wrong Note'. *Frieze* 119: 35–39. https://frieze.com/article/wrong-note.

Human Rights Watch, Arms Division. 1999. 'Acoustic Weapons: Memorandum for Convention on Conventional Weapons (CCW) Delegates'. First Annual Conference on CCW Amended Protocol II, United Nations, Geneva, 16 December. http://www.angelfire.com/nj4/hightechharassment/memo.htm.

Hyland, Julie. 2004. 'Britons Release Devastating Account of Torture and Abuse by U.S. Forces at Guantánamo'. *World Socialist Website*. 6 August. http://www.wsws.org/articles/2004/aug2004/guan-a06.shtml.

Ihde, Don. 2003. 'Auditory Imagination'. 61–66. Oxford: Berg.

James, William. 1884. 'What Is an Emotion?' *Mind* 9, no. 34: 188–205.

Jameson, Fredric. 1991. *Postmodernism, or, The Cultural Logic of Late Capitalism*. Durham, NC: Duke University Press.

———. 1995. *The Geopolitical Aesthetic: Cinema and Space in the World System*. Bloomington and London: Indiana University Press/BFI.

Jamieson, Ian. 1991. 'The Changing World of Work'. In *Rethinking Work Experience*, edited by A. Miller, A. G. Watts, and I. Jamieson, 55–84. London and Bristol, PA: Falmer Press.

Johnson, Bruce, and Martin Cloonan. 2009. *Dark Side of the Tune: Popular Music and Violence*. Farnham: Ashgate.

Kelley, Dean M. 1995. 'The Implosion of Mt. Carmel and Its Aftermath: Is It All Over Yet?' In *Armageddon in Waco: Critical Perspectives on the Branch Davidian Conflict*, 359–78. Chicago: University of Chicago Press.

Kittler, Friedrich. 1986. *Grammophon, Film, Typewriter*. Berlin: Brinkmann and Bose.

———. 1999a. *Gramophone, Film, Typewriter*. Translated by Geoffrey Winthrop-Young and Michael Wutz. Palo Alto, CA: Stanford University Press.

———. 1999b. *Hebbels Einbildungskraft—die dunkle Natur*. Frankfurt: Lang.

LaBelle, Brandon. 2010. *Acoustic Territories: Sound Culture and Everyday Life*. New York: Continuum.

Landsberger, Henry A. 1958. 'Hawthorne Revisited: Management and the Worker, Its Critics, and Developments in Human Relations in Industry'. *Cornell Studies in Industrial and Labor Relations* 9.

Lanza, Joseph. 2004. *Elevator Music: A Surreal History of Muzak, Easy-Listening, and Other Moodsong*. Revised and expanded edition. Ann Arbor: University of Michigan Press.

Leader, Anton, dir. 1963. *Children of the Damned*. UK: Metro-Goldwyn-Mayer.

Lee, Jennifer 8. 2001. 'An Audio Spotlight Creates a Personal Wall of Sound'. *New York Times*, 15 May. http://www.nytimes.com/2001/05/15/science/an-audio-spotlight-creates-a-personal-wall-of-sound.html.

Lefebvre, Henri. 1991. *The Production of Space*. Translated by Donald Nicholson-Smith. Oxford: Blackwell.

Levitin, Daniel J. 2006. *This Is Your Brain on Music*. New York: Dutton/Penguin.

Lewis, James R. 1995. 'Self-Fulfilling Stereotypes, the Anti-cult Movement and the Waco Confrontation'. In *Armageddon in Waco: Critical Perspectives on the Branch Davidian Conflict*, edited by Stuart A. Wright, 95–110. Chicago: University of Chicago Press.

Lowery, Oliver. 1992. 'Silent Subliminal Presentation System'. Patent application. United States Patent and Trademark Office. 27 October. http://patft.uspto.gov/netacgi/nph-Parser?Sect1=PTO1&Sect2=HITOFF&d=PALL&p=1&u=%2Fnetahtml%2FPTO%2Fsrchnum.htm&r=1&f=G&l=50&s1=5,159,703.PN.&OS=PN/5,159,703&RS=PN/5,159,703.

LRAD Corporation. 2009. 'HyperSonic Sound® Product Sheet'. Accessed 7 December. http://www.lradx.com/site/content/view/13/104.

Lukács, Georg. 2002. *History and Class Consciousness: Studies in Marxist Dialectics*. Translated by Rodney Livingstone. Cambridge, MA: MIT Press.

Madsen, Virginia. 2009. 'Cantata of Fire: Son et lumière in Waco Texas, Auscultation for a Shadow Play'. *Organised Sound* 14, no. 1: 89–99.

Manson, Charles. 1970. *Lie: The Love and Terror Cult*. Awareness Records. LP Record.

Mao, Jeff. 2018. 'Jalal Mansur Nuriddin, 1944–2018'. Red Bull Music Academy. https://one-last-note.redbullmusicacademy.com/2018/jalal-mansur-nuriddin.

Marx, Karl, and Friedrich Engels. 2007. *The Communist Manifesto*. Charleston, SC: BiblioBazaar.

Massachusetts Institute of Technology. 2018a. 'AlterEgo: Frequently Asked Questions'. MIT Media Lab. Accessed 21 May. https://www.media.mit.edu/projects/alterego/frequently-asked-questions/.

———. 2018b. 'Interfacing with Devices through Silent Speech'. MIT Media Lab. Accessed 21 May. https://www.media.mit.edu/videos/fi-alterego-2018-04-04/.

Maxwell, M. 1921. 'Mood Change Chart'. *The Edison Magazine* (January–February).

McLuhan, Marshall. 1962. *The Gutenberg Galaxy: The Making of Typographic Man*. Toronto: University of Toronto Press.

Mendes, Sam, dir. 2005. *Jarhead*. USA/Germany: Universal.

Merleau-Ponty, Maurice. 1962. *The Phenomenology of Perception*. Translated by Colin Smith. New York: Humanities Press.

Merrifield, Andrew. 2000. 'Henri Lefebvre: A Socialist in Space'. In *Thinking Space*, edited by Mike Crang and Nigel Thrift, 167–82. London: Routledge.

Milhaud, Darius. 1952. 'Lettre de Darius Milhaud'. *Revue musicale* 214 (June): 153.

Moore, Paul. 2003. 'Sectarian Sound and Cultural Identity in Northern Ireland'. In *The Auditory Culture Reader*, edited by Michael Bull and Les Back, 265–81. Oxford: Berg.

Moravec, Hans. 2000. *Robot: Mere Machine to Transcendent Mind*. Oxford: Oxford University Press.

Moving Sound Technologies. 2018. 'Mosquito MK4 with Multi-Age'. Accessed 21 November. http://www.movingsoundtech.com/our-products/mosquito-mk4-with-multi-age.

Naddaff, Ramona. 2009. 'No Blood, No Foul: Listening to Music at Guantánamo Bay'. Thinking Hearing: The Auditory Turn in the Humanities conference, University of Texas at Austin, 2 October.

Nattiez, Jean-Jacques. 1990. *Music and Discourse: Toward a Semiology of Music*. Translated by Carolyn Abbate. Princeton, NJ: Princeton University Press.

Negarestani, Reza. 2008. *Cyclonopedia: Complicity with Anonymous Materials*. Melbourne: re.press.
Noble, Ivan. 2002. 'Bug Sets Windows Shaking'. BBC News, 18 March. http://news.bbc.co.uk/1/hi/sci/tech/1879247.stm.
O'Callaghan, Casey. 2009a. 'Audition'. In *The Routledge Companion to the Philosophy of Psychology*, edited by John Symons and Paco Calvo, 579–91. New York: Routledge.
O'Dwyer, Rachel. 2009. 'Performing in the Acoustic Arena: The Role of the Portable Audio Device in a User's Experience of Urban Place'. In *Performing Technology: User Content and the New Digital Media*, edited by Franziska Schroeder, 97–106. Newcastle upon Tyne: Cambridge Scholars Publishing.
O'Neill, Brendan. 2010. 'Weaponizing Mozart: How Britain Is Using Classical Music as a Form of Social Control'. *Reason*, 24 February. https://reason.com/archives/2010/02/24/weoponizing-mozart.
Pasternak, Douglas. 1997. 'Wonder Weapons: The Pentagon's Quest for Nonlethal Arms Is Amazing. But Is It Smart?' *US News & World Report*, 29 June. https://www.bibliotecapleyades.net/sociopolitica/esp_sociopol_mindcon41.htm.
Peaches and Kim. 2007. *Escape*. Self-released. Compact disc.
Petersen, Julie K. 2007. *Understanding Surveillance Technologies: Spy Devices, Privacy, History, and Applications*. Revised and expanded 2nd edition. Boca Raton, FL: CRC Press.
Pieslak, Jonathan. 2009. *Sound Targets: American Soldiers and Music in the Iraq War*. Bloomington: Indiana University Press.
Porta, Giambattista della. 1558. *Magiae naturalis, sive, De miraculis rerum naturalium libri IIII*. Naples, Italy: Matthias Cancer.
Potter, Caroline, ed. 2016. *Erik Satie: Music, Art and Literature*. New York: Routledge.
Psywarrior. 2010. '193rd Special Operations Wing'. Accessed 30 May. http://www.psywarrior.com/ec130.html.
Pynchon, Thomas. 1975. *Gravity's Rainbow*. London: Picador.
Rand, Ayn. 1957. *Atlas Shrugged*. New York: Random House.
Rapp, Tobias. 2010. 'The Pain of Listening: Using Music as a Weapon at Guantanamo'. Translated by Christopher Sultan, *Spiegel Online International*, 15 January. http://www.spiegel.de/international/world/0,1518,672177,00.html.
Reality Games. 2004. *Kuma\War*. http://www.kumawar.com.
Reid, Tim. 2010. 'George W. Bush "Knew Guantánamo Prisoners Were Innocent"'. *The Times* (London), 9 April. Retrieved from https://www.globalpolicy.org/us-un-and-international-law-8-24/general-analysis-on-us-un-and-international-law/48931-george-w-bush-knew-guantanamo-prisoners-were-innocent.html.
Rejali, Darius. 2007. *Torture and Democracy*. Princeton, NJ: Princeton University Press.
Richardson, James T. 1995. 'Manufacturing Consent about Koresh: The Role of the Media in the Waco Tragedy'. In *Armageddon in Waco: Critical Perspectives on the Branch Davidian Conflict*, edited by Stuart A. Wright, 153–76. Chicago: University of Chicago Press.

Robbins, Thomas, and Dick Anthony. 1995. 'Sects and Violence'. In *Armageddon in Waco: Critical Perspectives on the Branch Davidian Conflict*, 236–59. Chicago: University of Chicago Press.

Rockstar Games. 1997. *Grand Theft Auto*. http://www.rockstargames.com/games.

Ronson, Jon. 2004. *The Men Who Stare at Goats*. London: Simon & Schuster.

Rorty, Amelie, and Martha Nussbaum, eds. 1992. *Essays on Aristotle's 'De Anima'*. Oxford: Oxford University Press.

Rose, Sherman A., dir. 1954. *Target Earth*. USA: Abtcon.

Rosenbaum, Art. 2007. *Art of Field Recording. Volume 1: 50 Years of Traditional American Music*. Dust-to-Digital DTD-08. Compact disc.

Rouse, Edward. 'PSYOP Equipment Systems'. *Psywarrior*. http://www.psywarrior.com/PSYOPEquipment.html.

Rubin, Rita. 2017. 'Alleged Acoustic Attack on US Diplomats Puzzling Experts'. *The Lancet*, 1 September. https://www.thelancet.com/journals/lancet/article/PIIS0140-6736(17)32359-0/fulltext.

Rushkoff, Douglas. 1996. *Media Virus! Hidden Agendas in Popular Culture*. New York: Ballantine Books.

Said, Edward. 1993. *Culture and Imperialism*. London: Vintage.

Sample, Ian. 2005. 'Read the Book, Seen the Movie? Now Smell It Too'. *The Guardian*, 7 April. http://www.guardian.co.uk/science/2005/apr/07/sciencenews.film.

Sassen, Saskia. 1998. *Globalization and Its Discontents*. New York: The New Press.

Scarry, Elaine. 1985. *The Body in Pain: The Making and Unmaking of the World*. New York: Oxford University Press.

Schaeffer, Pierre. 1966. *Traité des objets musicaux*. Paris: Éditions du Seuil.

Schafer, R. Murray. 1970. *The Book of Noise*. Vancouver: Price Print.

———. 1977. *The Tuning of the World*. New York: Random House.

———. 1993. *The Soundscape: Our Sonic Environment and the Tuning of the World*. Rochester, VT: Destiny Books.

———. 2003. 'Open Ears'. In *The Auditory Culture Reader*, edited by Michael Bull and Les Back, 25–39. Oxford: Berg.

Selfridge-Field, Eleanor. 1997. 'Experiments with Melody and Meter or the Effects of Music: The Edison-Bingham Music Research'. *The Musical Quarterly* 81, no. 2: 291–310.

Sella, Marshall. 2003. 'The Sound of Things to Come'. *New York Times*, 23 March. https://www.nytimes.com/2003/03/23/magazine/the-sound-of-things-to-come.html.

Serres, Michel. 1982. *The Parasite*. Translated by Lawrence A. Schehr. Baltimore: Johns Hopkins University Press.

Shakur, Tupac. 1997. *Hail Mary*. Death Row Records. 7" Record.

Shattuck, Roger. 1967. *The Banquet Years: The Origins of the Avant-Garde in France 1885 to World War I*. New York: Vintage Press.

Shupe, Anson, and Jeffrey K. Hadden. 1995. 'Cops, News Copy, and Public Opinion: Legitimacy and the Social; Construction of Evil in Waco'. In *Armageddon in Waco: Critical Perspectives on the Branch Davidian Conflict*, edited by Stuart A. Wright, 177–202. Chicago: University of Chicago Press.

Siegel, Daniel J. 2014. 'The Self Is Not Defined by the Boundaries of Our Skin'. *Psychology Today*, 28 February. https://www.psychologytoday.com/gb/blog/inspire-rewire/201402/the-self-is-not-defined-the-boundaries-our-skin.

Simmel, Georg. 1903. *Die Grosstädte und das Geistesleben*. Dresden: Petermann.

Skinner, B. F. 1938. *The Behavior of Organisms: An Experimental Analysis*. New York: Appleton-Century.

Sky News. 2018. 'US Embassy Worker in China Suffers "Brain Injury" amid "Sonic Attack" Fears'. 23 May. https://news.sky.com/story/us-embassy-worker-in-china-suffers-brain-injury-amid-sonic-attack-fears-11383119.

Smith, Colin. 2014. 'Scientists Explain in More Detail How We Hear via Bones in the Skull'. Imperial College London (website). 9 July. https://www.imperial.ac.uk/news/153374/scientists-explain-more-detail-hear-bones/.

Smith, F. J. 1979. 'Some Aspects of the Tritone and the Semitritone in the *Speculum Musicae*: The Non-emergence of the *Diabolus in Musica*'. *Journal of Musicological Research* 3, nos. 1–2: 63–74.

Smith, Mark M. 2003. 'Listening to the Heard Worlds of Antebellum America'. In *The Auditory Culture Reader*, edited by Michael Bull and Les Back, 137–64. Oxford: Berg.

Soja, Edward W. 1996. *Thirdspace: Journeys to Los Angeles and Other Real-and-Imagined Places*. Oxford: Blackwell.

Spielberg, Steven, dir. 2002. *Minority Report*. USA: Twentieth Century Fox.

Staver, M. 2009. 'He Writes the Rules That Make Their Eardrums Ring'. *Los Angeles Times*, 21 January. http://www.latimes.com/news/nationworld/nation/la-na-music-punishment21-2009jan21,0,1887999.story.

Steinmeyer, Jim. 1999. *The Science behind the Ghost: A Brief History of Pepper's Ghost*. Burbank, CA: Hahne.

Sterne, Jonathan. 2002. *The Audible Past: Cultural Origins of Sound Reproduction*. Durham, NC: Duke University Press.

Stone, Alan A. 1993. 'Report and Recommendations Concerning the Handling of Incidents Such as the Branch Davidian Standoff in Waco, Texas'. Washington, DC: US Department of Justice.

Storr, Anthony. 1992. *Notes for Music and the Mind*. New York: The Free Press.

Strauss, Erwin S. 1985. *How to Start Your Own Country*. Port Townsend, WA: Loompanics.

Tabor, James D. 1995. 'Religious Discourse and Failed Negotiations: The Dynamics of Biblical Apocalypticism'. In *Armageddon in Waco: Critical Perspectives on the Branch Davidian Conflict*, edited by Stuart A. Wright, 263–81. Chicago: University of Chicago Press.

Tandy, Vic, and Tony R. Lawrence. 1998. 'The Ghost in the Machine'. *Journal of the Society for Psychical Research* 62, no. 851: 360–64.

Taylor, Frederick Winslow. 1911. *The Principles of Scientific Management*. New York: Harper and Brothers.

Teyssot, Georges, et al. 1977. 'Heterotopias and the History of Spaces'. In *Il dispositivo Foucault*, edited by M. Cacciari et al., 23–36. Venice: CLUVA, Libreria Editrice.

Thacker, Eugene. 2004. 'Living Dead Networks'. *Fibreculture* 4. http://journal.fibreculture.org/issue4/issue4_thacker.html.

Tompkins, Dave. 2010. *How to Wreck a Nice Beach: The Vocoder from World War II to Hip-Hop; The Machine Speaks*. Brooklyn: StopSmiling/Melville House Publishing.

———. 2014. 'Future Shock'. *Wax Poetics*.

———. 2019. 'Keep Me in the Loop, You Dead Mechanism'. In AUDINT [Steve Goodman, Toby Heys, and Eleni Ikoniadou], eds. *Unsound: Undead*. Falmouth: Urbanomic x Sequence Press.

Trahair, Richard C. S. 1984. *The Humanist Temper: The Life and Work of Elton Mayo*. New Brunswick, NJ: Transaction Publishers.

Trower, Shelley. 2012. *Senses of Vibration: A History of the Pleasure and Pain of Sound*. London: Continuum.

Tsukamoto, Shin'ya, dir. 1989. *Tetsuo, the Iron Man*. Japan: JHV.

University of Michigan's Hatcher Library. 2010. 'Ghost Army Exhibition'. Accessed 21 May. http://www.lib.umich.edu/gallery/events/ghost-army.

Vassilatos, Gerry. 1996. 'The Sonic Doom of Vladimir Gavreau'. *Journal of Borderland Research* 52, no. 4 (October): 30–34. https://borderlandsciences.org/journal/vol/52/n04/Vassilatos_on_Vladimir_Gavreau.html.

Vinokur, Roman. 2004. 'Acoustic Noise as a Non-Lethal Weapon'. *Sound and Vibration* 38 (October): 19–23. https://pdfs.semanticscholar.org/4b37/8eef647a2382183507eadfd823a54c46fbb1.pdf.

Virilio, Paul. 1977. *Speed and Politics: An Essay on Dromology*. Translated by Mark Polizotti. New York: Semiotext(e).

———. 1994. *The Vision Machine*. Translated by Julie Rose. Bloomington: Indiana University Press.

———. 2002. *Desert Screen: War at the Speed of Light*. Translated by Michael Degener. London: Continuum.

———. 2006. *Speed and Politics*. Translated by Mark Polizzotti. Los Angeles: Semiotext(e).

Virilio, Paul, and Sylvère Lotringer. 1997. *Pure War*. Translated by Mark Polizzotti. New York: Semiotext(e).

Vokey, John R., and John D. Read. 1985. 'Subliminal Messages: Between the Devil and the Media'. *American Psychologist* 40, no. 11: 1231–39.

Walonick, David S. 1990. 'Effects of 6–10 Hz ELF on Brain Waves'. *Journal of Borderland Research* 46, nos. 3–4 (May–August): 32–36. https://borderlandsciences.org/journal/vol/46/n03-4/Walonick_Effects_6-10hz_ELF_on_Brain_Waves.html

Wark, McKenzie. 1994. *Virtual Geography: Living with Global Media Events*. Bloomington: Indiana University Press.

Weizman, Eyal. 2006. 'Israeli Military Using Post-Structuralism as "Operational Theory"'. *Infoshop News*. http://www.sholetteseminars.com/wp-content/uploads/2011/08/Israeli.MilitaryPostStruct.pdf.

The White House, Office of the Press Secretary. 2009. 'Presidential Memorandum: Closure of Detention Facilities at the Guantánamo Bay Naval Base'. The White House President Barack Obama, archives. https://obamawhitehouse.archives.gov/

the-press-office/presidential-memorandum-closure-dentention-facilities-guantanamo-bay-naval-base.

Wikipedia. 2018a. S.v. 'Category: Iraq War Films'. https://en.wikipedia.org/wiki/Category:Iraq_War_films.

———. 2018b. S.v. 'List of Iraq War Documentaries'. https://en.wikipedia.org/wiki/List_of_Iraq_War_documentaries.

Williams, Rhys H. 1995. 'Breaching the "Wall of Separation": The Balance between Religious Freedom and Social Order'. In *Armageddon in Waco: Critical Perspectives on the Branch Davidian Conflict*, edited by Stuart A. Wright, 299–322. Chicago: University of Chicago Press.

Winterbottom, Michael, and Mat Whitecross, creators. 2006. *Road to Guantánamo*. UK: Channel 4 Television Corp.

Womack, James P., Daniel T. Jones, and Daniel Roos. 1990. *The Machine that Changed the World*. New York: Rawson Associates.

Wouterloot, Lex. 1992. 'Silent Killing'. *Mediamatic* 6, no. 4: 251–55.

Wright, Steve. 2000.

Wright, Stuart A., ed. 1995. *Armageddon in Waco: Critical Perspectives on the Branch Davidian Conflict*. Chicago: University of Chicago Press.

Zarrella, Dan. 2008. 'My Viral Marketing Glossary'. 19 September. http://danzarrella.com/my-viral-marketing-glossary/.

The Zuckerman Institute at Columbia University. 2019. 'Engineers Translate Brain Signals Directly into Speech'. *Science Daily*. 29 January. https://www.sciencedaily.com/releases/2019/01/190129081919.htm.

Žižek, Slavoj. 2006. *The Parallax View*. Cambridge, MA: MIT Press.

OTHER SOURCES: BOOKS, JOURNALS, NEWS SOURCES AND CONFERENCE PROCEEDINGS

Adorno, Theodor. 1978. 'On the Fetish Character in Music and the Regression of Listening'. In *The Essential Frankfurt School Reader*, edited by Andrew Arato and Eike Gebhardt. Oxford: Blackwell.

Albright, Daniel, ed. 2004. *Modernism and Music: An Anthology of Sources*. Chicago: University of Chicago Press.

Anderson, Kevin. 1999. 'Koresh and the Waco Siege'. BBC News, 27 August. http://news.bbc.co.uk/1/hi/world/americas/431311.stm.

Aristotle. 1987. *De Anima (On the Soul)*. Translated by Hugh Lawson-Tancred. London: Penguin Classics.

Attlee, James. 2007. 'Towards Anarchitecture: Gordon Matta-Clark and Le Corbusier'. *Tate Papers* no. 7 (Spring). https://www.tate.org.uk/research/publications/tate-papers/07/towards-anarchitecture-gordon-matta-clark-and-le-corbusier.

Augoyard, Jean-François, and Henri Torgue, eds. 2005. *Sonic Experience: A Guide to Everyday Sounds*. Translated by Andra McCartney and David Paquette. Montreal: McGill–Queen's University Press.

Beard, William. 2001. *The Artist as Monster: The Cinema of David Cronenberg*. Toronto: University of Toronto Press.
Bentham, Jeremy. 1995. 'Panopticon (Preface)'. In *The Panopticon Writings*, edited by Miran Bozovic, 29–95. London: Verso.
Bingham, Nick, and Nigel Thrift. 2000. 'Some New Instructions for Travellers: The Geography of Bruno Latour and Michel Serres'. In *Thinking Space*, edited by Mike Crang and Nigel Thrift, 281–301. London: Routledge.
Brodeur, Paul. 1977. *The Zapping of America: Microwaves, Their Deadly Risk, and the Coverup*. New York: Norton.
Burke, Edmund. 1958. *A Philosophical Enquiry into the Origin of Our Ideas of the Sublime and Beautiful*. New York: Oxford University Press.
Calvino, Italo. 1972. *Le città invisibili*. Turin: Giulio Einaudi Editore.
Classen, Constance, ed. 2005. *The Book of Touch*. Oxford: Berg.
Corbin, Alain. 2003. 'The Auditory Markers of the Village'. In *The Auditory Culture Reader*, edited by Michael Bull and Les Back, 117–25. Oxford: Berg.
Deleuze, Gilles, and Félix Guattari. 1987. *A Thousand Plateaus: Capitalism and Schizophrenia*. Translated by Brian Massumi. Minneapolis: University of Minnesota Press.
DeLillo, Don. 1985. *White Noise*. New York: Viking.
Drobnick, Jim, ed. 2006. *The Smell Culture Reader*. Oxford: Berg.
Fackler, Guido. 1997. *'Des Lagers Stimme': Musik in den frühen Konzentrationslagern des NS-Regimes (1933–1936)*. PhD thesis, University of Freiburg, Germany.
Foucault, Michel. 1970. *The Order of Things*. Translated by Alan Sheridan. New York: Random House.
Gibbs, Samuel. 2018. 'Researchers Develop Device That Can "Hear" Your Internal Voice'. *The Guardian*, 6 April. https://www.theguardian.com/technology/2018/apr/06/researchers-develop-device-that-can-hear-your-internal-voice.
Gibson, William. 1984. *Neuromancer*. New York: Ace Books.
Gioia, Ted. 2006. *Work Songs*. Durham, NC: Duke University Press.
Ingold, Tim. 2008. 'Point, Line and Counterpoint: From Environment to Fluid Space'. *Neurobiology of 'Umwelt': How Living Beings Perceive the World*. IPSEN Foundation conference, Paris, France, 18 February.
Jones, Ernest. 1953. *The Life and Work of Sigmund Freud*, vol. 1. New York: Basic Books.
Justesen, Don R. 1975. 'Microwaves and Behavior'. *The American Psychologist* 30, no. 3: 390–401.
Kahn, Douglas. 2010. *Arts of the Spectrum: In the Nature of Electromagnetism*. Berkeley: University of California Press.
Kelion, Leo. 2013. 'Talking Train Window Adverts Tested by Sky Deutschland'. BBC News, 3 July. http://www.bbc.co.uk/news/technology-23167112.
Klein, Naomi. 2007. *The Shock Doctrine: The Rise of Disaster Capitalism*. New York: Metropolitan Books.
Lewis, James R. 2005. *Cults: A Reference Handbook*. Santa Barbara, CA: ABC-Clio.

McCoy, Alfred W. 2006. *A Question of Torture: CIA Interrogation, from the Cold War to the War on Terror*. New York: Henry Holt/Metropolitan Books.
McLuhan, Marshall. 2004. *Understanding Me: Lectures and Interviews*, edited by Stephanie McLuhan and David Staine. Cambridge, MA: MIT Press.
McLuhan, Marshall, Quentin Fiore, and Jerome Agel. 1967. *The Medium Is the Massage: An Inventory of Effects*. New York: Random House.
Meek, James. 2005. 'Nobody Is Talking'. *The Guardian*, 18 February. http://www.guardian.co.uk/world/2005/feb/18/usa.afghanistan.
Merleau-Ponty, Maurice. 1968. *The Visible and the Invisible*. Translated by Alphonso Lingis. Evanston, IL: Northwestern University Press.
Nietzsche, Friedrich. 1966. *Werke in drei Bänden*. Edited by Karl Schlechta. Munich: Carl Hanser Verlag.
———. 1967. *The Birth of Tragedy out of the Spirit of Music and the Case of Wagner*. Translated by Walter Kaufmann. New York: Vintage Books.
Nudds, Matthew. 2001. 'Experiencing the Production of Sounds'. *European Journal of Philosophy* 9, no. 2: 210–29.
O'Callaghan, Casey. 2009b. 'Constructing a Theory of Sounds'. *Oxford Studies in Metaphysics* 5: 247–70.
———. 2009c. 'Is Speech Special?' University of British Columbia Working Papers in Linguistics, Interlocution Workshop Proceedings 24 (July): 57–64.
Omega Foundation. 2000. 'Crowd Control Technologies: An Assessment of Crowd Control Technology Options for the European Union' (EP/1/IV/B/STOA/99/14/01). Presented to the LIBE Committee of the European Parliament, 29 August.
Philo, Chris. 2000. 'Foucault's Geography'. In *Thinking Space*, edited by Michael Crang and Nigel Thrift, 205–38. London: Routledge.
Plato. 1974. *The Republic*. New York: Penguin.
Pratella, Francesco Balilla. 1912. *Manifesto of Futurist Musicians* (*Musica futurista di Balilla Pratella*). n.p.
Rodwell, R. 1973. 'Squawk Box Technology'. *New Scientist* 59, no. 864: 667–68.
Roethlisberger, Fritz J., and William J. Dickson. 1939. *Management and the Worker: An Account of a Research Program Conducted by the Western Electric Company, Hawthorne Works, Chicago*. Cambridge, MA: Harvard University Press.
Rose, David. 2004. 'Revealed: The Full Story of the Guantánamo Britons'. *The Observer*, 13 March. http://www.guardian.co.uk/uk/2004/mar/14/terrorism.guantanamo.
Schopenhauer, Arthur. 1969. *The World as Will and Representation*, 2nd ed. Translated by E. F. J. Payne. New York: Dover.
Shannon, Claude, and William Weaver. 1949. *The Mathematical Theory of Communication*. Urbana: University of Illinois Press.
Skinner, B. F. 1953. *Science and Human Behavior*. New York: Macmillan.
Smith, Adam. 1776. *An Inquiry into the Nature and Causes of the Wealth of Nations*. London: W. Strahan and T. Cadell.
Soja, Edward. 1989. *Postmodern Geographies: The Reassertion of Space in Critical Social Theory*. London: Verso.

Stone, Geoffrey R. 2004. *Perilous Times: Free Speech in Wartime from the Sedition Act of 1798 to the War on Terrorism*. London: W. W. Norton and Company.
Strawson, P. F. 1959. *Individuals: An Essay in Descriptive Metaphysics*. New York: Routledge.
Thomas, Gordon. 1989. *Journey into Madness: The True Story of Secret CIA Mind Control and Medical Abuse*. New York: Bantam Books.
Thompson, Emily. 2004. *The Soundscape of Modernity: Architectural Acoustics and the Culture of Listening in America, 1900–1933*. Cambridge, MA: MIT Press.
Virilo, Paul. 1991. *The Aesthetics of Disappearance*. Translated by Phil Beitchman. New York: Semiotext(e).
Webster, D. 2005. 'The Man in the Hood and New Accounts of Prisoner Abuse in Iraq'. *Vanity Fair*, 1 February.
Wegner, Philip E. 2006. 'Periodizing Jameson, or, Notes toward a Cultural Logic of Globalization'. In *On Jameson: From Postmodernism to Globalization*, edited by Caren Irr and Ian Buchanan, 241–79. Albany: State University of New York Press.
Yerkes, Robert M., and John D. Dodson. 1908. 'The Relation of Strength of Stimulus to Rapidity of Habit-Formation'. *Journal of Comparative Neurology and Psychology* 18: 459–82.

OTHER SOURCES: FILMOGRAPHY

Burton, Tim, dir. 1996. *Mars Attacks*. USA: Warner Bros.
Coppola, Francis Ford, dir. 1979. *Apocalypse Now*. USA: Omni Zoetrope.
Grayson, Godfrey, dir. 1949. *Dick Barton Strikes Back*. UK: Hammer Films.
Kubrick, Stanley, dir. 1971. *A Clockwork Orange*. UK/USA: Polaris Productions.
Lynch, David, dir. 1984. *Dune*. USA: Dino De Laurentiis Corporation.
Martino, Sergio, dir. 1983. *2019: After the Fall of New York*. Italy/France: Medusa Film.
Proyas, Alex, dir. 1998. *Dark City*. Australia/USA: Mystery Clock Cinema.
Saleh, Tarik, dir. 2010. *Metropia*. Sweden/Denmark/Norway/Finland: Atmo.
Sears, Fred F., dir. 1956. *Earth vs. the Flying Saucers*. USA: Columbia.

OTHER SOURCES: SPECIFIC WEBOGRAPHY

Animal Voice. 2009. 'Infrasonic Animal Communication'. Accessed 23 February. http://www.animalvoice.com/seismic.htm.
CCTV Direct. 2010. 'Mosquito Mk4 Ultrasonic Youth Deterrent'. Accessed 16 May. http://www.cctvdirect.co.uk/products/Mosquito-Mk4-Ultrasonic-Youth-Detterent.html.
Encyclopaedia Britannica. 2018. S.v. 'Transmission of Sound by Bone Conduction'. https://www.britannica.com/science/ear/Transmission-of-sound-by-bone-conduction.

Esri India. 2018. '"Sensors" for Smart "Cities"'. Accessed 2 March. http://www.esri.in/esri-news/publication/vol9-issue1/articles/sensors-for-smart-cities.

Flat Screen Support. 2012. 'Feonic, Terfenol, SFX Technology, MTV Wowee One, Gel Audio, Clearsound, NTX Technologies—Distributed Mode Loudspeakers DML'. 31 January. https://flatscreensupport.wordpress.com/2012/01/31/feonic-terfenol-sfx-technology-mtv-wowee-one-gel-audio-clearsound-ntx-technologies-distributed-mode-loudspeakers-dml/.

Global Security. 2010. 'Desert Storm Statistics and Information'. Accessed 14 May. http://www.globalsecurity.org/military/ops/desert_storm.htm.

HAARP [High-Frequency Active Auroral Research Program]. 2010. 'HAARP Executive Summary'. Accessed 14 May. http://www.bariumblues.com/haarp_executive_summary.htm.

Holosonic Research Labs. 2009. 'Customer List'. Accessed 16 November. http://www.holosonics.com/applications.html.

Muzak. 2008. 'Audio Architecture'. Accessed 3 July. http://music.muzak.com/why_muzak/.

Pieslak, Jonathan. 2010. 'CJ Grisham'. Interviews with American soldiers about the use of music during war. Accessed 4 May. http://jon.pieslak.com/asom/CJGrisham.htm.

Project Guttenberg. 2009. 'My Life and Work by Henry Ford'. Accessed 4 January. http://www.gutenberg.net/etext/7213.

Psywarrior. 2018. 'History of PsyOps'. Accessed 23 April. http://www.psywarrior.com/psyhist.html.

Raven1.net. 2010. 'Conspiracy Theories about Frequency-Based Weapons'. Accessed 21 May http://www.raven1.net.

Reprieve. 2018. 'Guantánamo Bay'. Accessed 29 December. https://reprieve.org.uk/topic/guantanamo-bay/.

Russolo, Luigi. 'The Art of Noises'. *The Niuean Pop Cultural Archive.* http://www.unknown.nu/futurism/noises.html. Accessed April 4, 2008.

Society for Ethnomusicology. 2010. 'Position Statement on Torture'. Accessed 26 April. http://webdb.iu.edu/sem/scripts/aboutus/aboutsem/positionstatements/position_statement_torture.cfm.

Something Awful. 2010. 'Iraq War Albums by U.S. Soldiers'. Accessed 27 April. http://www.somethingawful.com/d/news/iraq-war-albums.php?page=2.

Spacedog. 2008. 'Infrasonic Psychological Experiment Conducted in Liverpool Metropolitan Cathedral by Sarah Angliss'. Accessed 7 November. http://spacedog.biz/sonicart/infrasonic.

———. 2008. 'Who We Are'. Accessed 12 March. http://www.spacedog.biz/extras/Infrasonic/whoweare.htm.

Tandy, Vic. 2010. 'The Ghost in the Machine'. Accessed 25 May. http://www.psy.herts.ac.uk/ghost/ghost-in-machine.pdf.

Timms, Dominic. 2003. 'Iraq War Game Comes under Fire'. *The Guardian*, 15 August. http://www.guardian.co.uk/media/2003/aug/15/digitalmedia.games.

United States Institute of Peace. 2002. 'Report of the Chilean National Commission on Truth and Reconciliation'. 4 October. https://www.usip.org/sites/default/files/resources/collections/truth_commissions/Chile90-Report/Chile90-Report.pdf.

University of Alaska Fairbanks, Geophysical Institute. 2009. 'Measuring Naturally Occurring and Man-Made Infrasound at the Geophysical Institute of the University of Alaska Fairbanks'. Accessed 19 February. http://www.gi.alaska.edu/infrasound/.

———. 2010. 'High-Frequency Active Auroral Research Program (HAARP)'. Accessed 14 May. http://www.haarp.alaska.edu/haarp/gen.html.

Webster, Donovan. 2005. 'The Man in the Hood: And New Accounts of Prisoner Abuse in Iraq'. *Vanity Fair*. February.

World Federation of Music Therapy. 2010. '2011 World Congress'. Accessed 25 May. http://www.wfmt.info/WFMT/2011_World_Congress.html.

OTHER SOURCES: WEBSITES

Amnesty International. 2010. Accessed 11 September. http://www.amnesty.org/.
Amnesty USA. 2009. Accessed 19 March. http://www.amnestyusa.org.
Artesonoro.org. 2008. Accessed 12 March. http://www.artesonoro.org/sonicweapons/.
ATCSD.com. 2009. Accessed 23 February. http://www.atcsd.com/site/.
Cadre Media Lab. 2008. Accessed 4 April. http://cadre.sjsu.edu/switch/sound/articles/wendt/folder6/ng632.htm.
Dust to Digital. 2009. Accessed 10 April. http://dust-digital.com/.
Feonic. 2018. Accessed 2 March. http://www.feonic.com/.
Human Rights Watch. 2009. Accessed 11 September. http://www.hrw.org/.
Internet Movie Database. 2010. Accessed 25 April. http://www.imdb.com/.
LRAD Corporation. 2009. Accessed 23 February. https://www.lradx.com.
Mood:Media. 2008. Accessed 3 July. http://music.muzak.com.
Psywarrior. 2009. Accessed 23 April. http://www.psywarrior.com/.
7Hz.org. 2008. Accessed 4 April. http://www.7hz.org/s_arford/infrasound.html.
23five Incorporated. 2008. Accessed 12 March. http://www.23five.org/infrasound/.
World Socialist Web Site. 2010. Accessed April 23. http://www.wsws.org.
Wowee One. 2018. Accessed 2 March. http://www.woweeone.com/.

Index

4th25, 88

Abu Ghraib, 55, 100, 110, 169
AC/DC, 21n25
acoustics, 2, 5, 18n9, 32, 36–37, 50, 54, 58, 64, 71–74, 77, 79, 83–86, 90–92, 94–96, 98, 100–101, 103, 111–14, 124–26, 130, 136, 139, 152, 157–59, 163, 169; acoustic attacks, xi, 2, 10, 15, 177–80; electroacoustics, 137, 152
Adorno, Theodor, 104, 107
agency, x, 3–4, 8–9, 11, 14, 18n8, 20n20, 25, 35, 44, 54, 65, 70, 75n4, 96, 98, 100, 109, 116, 129, 133, 137–39, 141, 152–53, 160, 162, 166–67, 170, 173, 175, 183; liquid, 184–85; sensorial, 3; social, 11; ultrasonic, 7–8
agriculture, 23, 42, 126
Ahmed, Ruhal, 79
AI. *See* artificial intelligence.
Alexander, John, 129
Al-Harith, Jamal Udeen, 79, 119n4
Al-Mallah, Haitham, 81, 116
alienation, 4, 42, 53, 62, 68, 87, 142
AlterEgo, 177, 180–83
Altmann, Jürgen, 131, 178
animism, 161, 170 72

Anthony, Dick, 63
antisocial behaviour, 86
Aristotle, 3
Armageddon, 68
artificial intelligence (AI), xi, 2, 149n4, 160, 177, 181–82, 184; audio intelligence, 2, 34, 130, 173, 175, 182
Attali, Jacques, 8, 57–58, 67, 71, 80, 97, 102, 108–9, 141
AUDINT, 173
audiotopia, 30, 70–72
Augé, Marc, 114
avant-garde, 57, 79–81

Babbage, Charles, 24
Bach, Johann Sebastian, 85
Back, Les, 96–97
backmasking, 16, 20n24
Ballard, J. G., 71–72, 75, 147–48
Bassett, Caroline, 160, 166
Bateson, Gregory, 6
Baudrillard, Jean, 112
Bayoumi, Moustafa, 81, 84, 105
becoming, 173–74
Beethoven, Ludwig van, x, 85
Bell, Alexander Graham, 120n9, 137
Benjamin, Walter, 147
Bentham, Jeremy, 39, 47n9

Benton, Stevie, 85
Benton, William, 26
Bhabha, Homi K., 20
Bijsterveld, Karin, 163
Bingham, Walter Van Dyke, 36–38
Bishop, Jack, 137
Bissell, David, 173
Bobb, Mitchell, 135
body, x, 1–2, 4–7, 14, 16, 18n11, 19n17, 23, 28–31, 37–39, 44–45, 47n10, 47n11, 52, 74, 81–82, 89–93, 97–99, 115, 118, 120n7, 124, 126–27, 132, 137–38, 148n1, 152–54, 159, 163, 165–69, 171, 181–82; antenna-body, 4–7, 18n5, 44–45, 53, 90, 139, 147, 164, 167–68, 175, 182, 185; body without organs, 168; databody, 161, 181; disembodiment, xi, 74, 90–92, 107, 120n7, 120n9, 137, 143, 175; docile, 29; embodiment, 4, 6, 8, 17n3, 19n17, 30, 43, 45, 60, 90–93, 106–7, 118, 162, 167, 186n2; habitual, 182, 186n3; holobody, 5; industrialised, 23–26, 28–29, 32, 35, 44, 156; lived, 174, 175n1; military, 133; of noise, 49; politic, 23, 97; social, mass, 9, 23, 26, 32–33, 36, 40–41, 59, 75n4, 89, 92–93, 124, 137, 146; sonic, 90; waveformed, 14, 28, 124, 132; unheard, 132–33
breakdancing, 156
Brown, James, x
Brown, Mick, 111
Buddhism, 54, 155–56
Bull, Michael, 96–97
Burgess, Anthony, 86
Burris-Meyer, Harold, 40, 47n12
Burroughs, William, 74
Bush, George H. W., 110
Bush, George W., 110

Cahn, Edward L., 128
Cameron, Donald Ewen, 104–5
Camolez, José Miguel, 82

Cannon-Bard theory of emotion, 39, 47n11
capitalism, 9, 24, 31, 33–34, 43, 46n6, 108, 125, 158–59, 170, 172–73; capital, 20, 27–28, 31, 33, 41, 43, 158
capoeira, 80, 156
Cardinell, Richard L., 40, 47n12
Carterby, Ben, 135
cartography, 3, 16, 39, 46n4, 57, 159, 185; sonic, 5, 9, 30, 67, 119, 130, 167
Central Intelligence Agency (CIA), 83, 104–5
chaos, 20n20, 64–66, 75n4, 93, 95–97, 106, 110, 138
Chatwin, Bruce, 67–68
Chion, Michael, 137, 186n1
Christianity, 16, 68–70, 93–94, 148; Christian rock, 59–60, 69; Christian rap, 69
CIA. *See* Central Intelligence Agency.
Cixous, Hélène, 20n20
Cloonan, Martin, 94, 100
Cocteau, Jean, 155
Coetzee, J. M., 68
Cole, Ronald, 54
communism, 24, 136
Connor, Steven, 20n21
Conroy, John, 82
consciousness, 7, 39, 93, 107, 145, 149n1, 165–68, 183–84; collective, 79, 143, 182; public, 182; Siamese, 133–34; waveform, 138;
Coppola, Francis Ford, 174
Corbin, Alain, 169–70
Cosmides, Leda, 104
country music, 8, 79–80
Crary, Jonathan, 146–47
Crocker, Stephen, 144–45
Cronenberg, David: *Videodrome*, 149n3; *Scanners*, 183
cults, 60–61, 63, 72, 75n4

Cusick, Suzanne, 81, 83, 89, 105, 120n7

dancehall, 70
death, 18n11, 42, 45, 49, 52, 55, 57, 60, 71, 73, 78, 92, 97, 108–9, 127–28, 141, 174; sonic, 108
De Certeau, Michel, 5, 17n2, 18n8
Debord, Guy, 18n4, 55
DeLanda, Manuel, 23–24, 28, 41, 108
Deleuze, Gilles, 8, 20n23, 55, 168
détournement, 55, 87
devil, 17, 69–70, 93–94
Dickens, Charles, 171
Dircks, Henry, 171
disco, xi, 8, 81–87, 116
DIY music, 80
DJ Smokey, ix–x
Dr Dre, 171
duration, 11, 36, 54, 57, 65
Durkheim, Émile, 47n8
Durrett, Deanne, 135
Dyer, Richard, 81

ecology, sonic, 156, 167
Edison, Thomas, 36–38, 46n7, 108, 127, 137
Electric Light Orchestra (ELO), 20n24
Eminem, 20n24, 79
enclosure, 116, 128
Engels, Friedrich, 31, 46n6
evolution, 156, 157
excess, xi, 70–72, 81, 95–96, 99, 103, 106, 108, 110, 114–15, 120n10, 133–34, 174–75; management, 174; removal of, 164, 175
exorcism, 91–94

fatigue curve, 40, 47n12
Faucher, Léon, 28
Federal Bureau of Investigation (FBI), 1, 13, 49, 51–55, 57–59, 62–68, 70–72, 74
Feonic, x, 151–53, 172
fiction, sonic, 148, 155, 177

footwork, 156
Ford, Henry, 24–25, 27, 29, 31; Fordism, 23–25, 27, 31, 33, 35, 58, 59, 66, 154
Foucault, Michel, 4, 6, 10, 16, 18n5, 18n6, 23, 26–30, 33–34, 43, 46n4, 72, 77, 91, 93, 106–108, 110–12, 116
Fuller, Matthew, 97, 133, 163

Gaffney, Edward Jr., 50, 52
Galanter, Marc, 63
Gavreau, Vladimir, 128
geography, 5, 9–13, 19n18, 44, 46n5, 130, 141–44; psychogeography, 9–10, 18n4, 137–39
Gerard, Philip, 135
Ghost Army, 102, 135, 166
Gierke, Henning von, 127
God, 33–34, 53, 68, 148
Goldberg, Ken, 143
Goodman, Steve, 17n3, 88, 97, 125–26, 131, 139, 147, 168, 170
Gore, Tipper, 60
Gorriors of Ragnarok, 88
Gramsci, Antonio, 24–25
GrandMixer DXT, ix–x
Grand Theft Auto, 87
Guantánamo Bay, 1, 7–8, 13, 41, 55, 65, 77–79, 82–87, 95, 100, 105, 107, 118, 119n1, 119n2, 119n4, 120n6, 139–40, 148, 169
Guattari, Félix, 8, 20n23, 55
Gulf War, 112, 128, 135–36

Hadden, J. K., 54, 57, 61
Hall, Josh, 61
Hallett, Mark, 179
hallucination, 15, 79, 120n13, 127, 143, 149n3, 153, 171, 185
harmony, 19n17, 27–28, 32, 35, 41, 65, 69–70, 87, 89, 102, 115, 143, 170; disharmony, 42
Harris, Harry, 78

heavy metal, 8, 79, 93, 94, 106; nu metal, 85
Hebdige, Dick, 85
Hennlich, Andrew, 55
Henriques, Julian, 11–12, 70, 90–92, 114, 120n9
Hertz, Heinrich, 7, 18n10; hertz (hz), x, 73, 127, 148n1, 164–65
heterology, 46n4
heterotopia, 10, 30, 46n4, 72
Hitchcock, Alfred, 128
Hoffman, Rick, 78
homo sacer, 87
house music, 156
hooks, bell, 10
Howard, John, 94–95
HSS. *See* HyperSonic Sound System
Hurt Locker, The, 88
HyperSonic Sound System, 2, 6, 9, 13, 17, 123–26, 130, 132–34, 137–48, 152–53, 164, 167–68, 185

identity, 4, 20n20, 30, 46n4, 56, 62, 64, 80, 89, 105, 107, 109, 114, 116, 129, 145, 157, 182, 185
ideology, 7–8, 39, 140, 147
Industrialisation, 17n1, 23–25, 28–29, 31–32, 39, 44–45, 46n2, 70, 156; Industrial Revolution, 45, 45n1; Second Industrial Revolution, 9, 23
infrasound, infrasonic, 1, 10–11, 14, 18n7, 32, 71, 127–28, 178
intensity, 19n16, 39, 82–84, 101, 108, 126–27, 130, 143, 156, 185; sonic, 78, 87, 101
interface, 6, 44, 91, 153, 159–60, 162, 165–66, 172, 175, 177, 181
internet, 12, 26, 139, 143, 163, 166–67, 181, 184–85; of things, 166
Islam, 94, 106, 121n14; Muslim detainees, 78, 81, 87, 109–10

Jacob, Max, 155
James, William 39, 47n10

Jameson, Fredric, 7, 12, 55, 103, 182, 185
Jarhead, 87
Johnson, Bruce, 94, 100
Jones, Jim, 61
Jones, Peter, 153, 172
Judas Priest, 55

Kapur, Arnav, 182
Kelley, Dean M., 62
Kerry, John, 110
Kittler, Friedrich, 12, 120n9, 174
Kool Herc, ix
Koresh, David, 50–54, 59–60, 62–64, 69–70, 74; Branch Davidians, 1, 49–50, 52, 54, 57, 60, 65–66, 71, 183

LaBelle, Brandon, 151
labour, 20, 23–25, 29, 33, 40, 43–44, 143, 158
Lange, Carl, 39, 47n10
Lanza, Joseph, 27–29, 33, 42, 45, 46n5, 154
Latour, Bruno, 173
Lee, Jennifer 8., 125
Lefebvre, Henri, 5, 10, 12, 19n17, 19n19, 180
Lethbridge, Thomas Charles, 171
Lewis, James, 61
living dead, 13–14, 16, 72–75, 94–95, 107, 109. *See also* undead; zombie
Logan, John A., 103
Long Range Acoustic Device (LRAD), 9, 18n13, 84, 123, 130
Lowery, Oliver, 130
LRAD. *See* Long Range Acoustic Device.
Lukács, Georg, 30

Madsen, Virginia, 50, 57, 65, 74
Manson, Charles, 20n24, 60
Manson, Marilyn, 60
Marx, Karl, 31, 46n6
Massachusetts Institute of Technology (MIT), 125, 177, 181–83

Matta-Clark, Gordon, 55
Maxwell, William, 38
Mayo, Elton, 47n8
McCarthyism, 93
McConnell, Mitch, 179
McLuhan, Marshal, 46n5, 125, 144
melody, 28, 32, 37, 68, 148
Merleau-Ponty, Maurice, 174, 175n1, 182, 186n3
Metallica, 79
military–entertainment complex, 1, 8–9, 41, 80, 86, 87–89, 103, 108–9, 115–18, 133, 139–40
military–industrial complex, 1, 8, 28, 41, 77
MIT. *See* Massachusetts Institute of Technology.
MKUltra, 104–5
Moore, Lawrence, 62
Moore, Paul, 137
Moravec, Hans, 7
Mosquito, 9, 19n14, 131–32
Mount Carmel, 49, 51–52, 58, 60, 64, 66–68, 71–72, 74
Mozart, Wolfgang Amadeus, 86
Muzak, 13, 17, 23–45, 46n3, 46n5, 47n12, 66, 153–54, 156, 158, 183

Naddaff, Ramona, 89, 105
Nattiez, Jean-Jacques, 96
Nauert, Heather, 178
Naveh, Shimon, 55
Negarestani, Reza, 170, 173
Noble, Ivan, 152
noise, 4, 6, 10–11, 18n13, 19n15, 19n20, 28, 32, 45, 50, 54, 57–58, 64, 66–67, 70–72, 79, 82–84, 94–97, 100–102, 104, 107–8, 112, 115, 120n8, 120n10, 129, 140–41, 144–45, 148, 153, 155, 157, 163–64, 173–74, 178, 180
Noriega, Manuel, 55
Norris, Elwood G., 124

Obama, Barack, 78, 178

occult, 72, 172
O'Neill, Brendan, 86
O'Neill, Don, 40
O'Reilly, Bill, 85
oscillation, 11, 44, 89, 96, 101, 148n1, 153, 165–66, 173, 185; oscillating subject, 4, 18n5
outsider, 61–62, 69–70, 75n4

pain, 3, 28–29, 42, 79, 88, 90, 92, 97–100, 102, 104, 108, 113, 115, 117–18, 120n7, 120n8, 127–33
Pareto, Vilfredo, 47n8
Payne, Donald, 81
Peaches and Kim, 88
Peignot, Jérôme, 137
Pepper, John, 171
perception, 2–3, 5–7, 10–11, 15–16, 17n3, 44, 47n10, 65, 67, 71, 90, 92, 101, 118, 120n13, 125–26, 130, 133–34, 136–38, 140–41, 146, 156, 160, 165, 171, 175n1, 182, 184, 186n3
Pieslak, Jonathan, 106–7
Plato, 106
pleasure, 1, 10, 102, 117, 148, 158, 169, 184–85
poiesis, 106
Pompei, Joseph, 125, 178
Pompeo, Mike, 179
pop music, 8, 20n24, 63, 69, 72, 79, 82, 88, 111, 157
Porta, Giambattista della, 171
Potter, Caroline, 155
power, 6–7, 9, 29–30, 60, 65–66, 103, 107, 120n7, 141, 163; divine, 148; economy of, 110; of frequencies, 14, 16; of music, 36–37, 59, 69; music as a tool of, 102; relations, 4, 18n5, 23, 28, 34–35, 39, 46, 89–90, 101; state, 86; structures of, 111; ultrasonic, 126; weapons of, 57
pressure, 1–7, 9–11, 17, 35, 45, 52, 66, 71, 90, 93, 95, 98, 100, 102–5,

112, 114–15, 117, 119n2, 127–28, 133–34, 138, 141, 159, 175, 177, 179, 183–85
Price, Henry Habberley, 171
psyops, 72–74, 75n1, 136–37
Pynchon, Thomas, 75n4
Pythagoras, 169

Ramirez, Richard, 16, 21n25
Rand, Ayn, 128
rap, 79–80, 83, 87, 109, 156, 170–71; gangsta rap, 87; 'rapparitions', 170
Rapp, Tobias, 79
Rasul, Shafiq, 78–79, 119n4
Reichenbach, Tobias, 165
Reid, Tim, 78
Rejali, Darius, 82–83, 120n6
Reno, Janet, 49
repetition, 1, 8, 18n12, 25–26, 28–30, 33, 35–37, 39, 41–43, 54, 59, 64–65, 77, 84, 89–90, 92–93, 96, 101, 105, 107–8, 110, 118, 133, 154
rhythm, 9, 11, 13, 15, 23, 25–28, 30–31, 33, 35, 38, 40–41, 44–45, 65, 79, 87, 89, 93–94, 96, 99, 100, 103, 107, 116, 132–33, 136, 142–43, 147–48, 154, 156, 183; arrhythmia, 139; biorhythms, 28, 127; microrhythms, 101; rhythmanalysis, 10, 19n17
Richardson, James, 52–53, 64
ritual, 64–65, 70, 94, 99, 102, 110, 114, 154–55
Robbins, Thomas, 63
rock music, 55, 60, 82, 109, 111
Rodríguez, Bruno, 178
Ronson, Jon, 79
Rose, Sherman A., 128
Rushkoff, Douglas, 20n22
Russolo, Luigi, 19n15

Sacco, Paul, 83
Sachnowitz, Herman, 135
Said, Edward, 13, 142
Sassen, Saskia, 31–32, 117–18

Satie, Erik, 153–58
Scarry, Elaine, 98–100, 102, 115–16
Schaeffer, Pierre, 137, 169, 186n1
Schafer, R. Murray, 18n9, 63, 66–67, 70–71, 97, 101, 120n10, 137–38, 148
schizophrenia, 16, 20n23, 103–5, 120n11, 120n12, 125, 133; schizoid, 103–104, 120n11;
schizophonia, 137–138
Schmid, Charles, 60
self, ix, 12, 18n5, 60, 90, 92, 98–101, 107, 111, 130, 133–34, 138, 143, 181; self-harm, self-immolation, self-mutilation, 78, 105, 120n7, 156; self-stigmatisation, 62; self-surveillance, 34, 47n9; selfie, 160
Selfridge-Field, Eleanor, 37–38
Sella, Marshall, 124
Serres, Michel, 6, 19n20, 20n21, 144–45, 182
Shakur, Tupac, 171
Shannon, Claude, 145, 149n4
Shupe, Ansen, 54, 61
silence, ix, xi, 4, 8, 14, 25, 27, 40, 44–45, 50, 52, 65, 67, 71, 75, 91, 93, 94–95, 98–100, 102, 108, 115, 118, 120n10, 129, 138, 140–41, 144, 145–46, 148, 154, 157, 161, 164, 174; Silent Sound Spread Spectrum, 130; silent speech, 181; silent system, 94
Simmel, Georg, 3, 142–43
Sinatra, Nancy, 54, 64
Skinner, Burrhus Frederic, 58–59
sleep, 148n1, 164–66, 169, 185; sleep deprivation, 8, 54, 83, 119n1, 119n2
Smith, Mark, 94
Snoop Dogg, 171
sociosonic, 72
Soja, Edward, 10, 19n18
spatiality, sonic, 11–12, 25, 27–28, 30, 32–34, 40–45, 49, 62, 67, 70, 72–74, 106, 113–16, 153

speakers, speaker systems, ix–xi, 1–2, 13, 27, 34, 36, 39, 43, 44, 53–54, 55, 63, 65, 66–67, 71, 73–74, 82, 93, 95, 102, 112, 123, 124, 130, 153, 154, 158–59, 163, 164, 172, 175; sound systems, ix–xi, 1–2, 4, 6, 13, 27, 36, 44, 49, 63, 65–66, 70, 95, 113, 152–53, 159, 164, 167–70, 172
Spears, Britney, 20n24
Spector, Phil, 111
Spinoza, Baruch, 6
Spotify, 159
Squier, George Owen, 26–27, 31, 35, 43, 46n3, 46n5
Staver, Matthew, 83
Sterne, Jonathan, 3, 186n2
stone tape theory, 171–73
Storr, Anthony, 67–69
strategy, sonic 1, 41, 49, 57, 72, 112
subjectivity, 4, 6, 9, 26, 103, 138, 142–43
subsonic, 64
surrealism, 56
surveillance, sonic, 51

Taylor, Frederick Winslow, 23–25, 45n2; Taylorism, 24, 42–43
teenagers, 19n14, 132
temporality, 12, 29, 38, 43–44, 130, 141, 158, 169
Terfenol-D, x–xi, 152
texture, sonic, 19n15
thirding, 10, 19n20
Thomas, Gordon, 104
time, sonic, 11
Tompkins, Dave, 172
Tooby, John, 104
torture, sonic, 1, 7, 59, 77, 79, 82–86, 93–96, 99–100, 104–9, 114–16, 118, 119n1, 140,
Trower, Shelley, 167–68, 173
Trump, Donald, x, 78, 179
Turner, Victor, 70

ultrasonic, 1, 7–11, 14, 17, 18n7, 19n14, 71, 123–26, 130–34, 137–48, 178. *See also* infrasonic
undead, 107, 172–75
unsound, 2, 5, 13, 17, 17n3, 74, 93, 119, 127–29, 138–39, 145, 147, 161, 166, 174–75, 179–80, 183–85; Unsound System, 177–80, 185

Van Halen, 55
Verdi, 83
vibration, ix, 17n3, 57, 63, 65, 74, 89–90, 126–27, 130, 136, 140–41, 151–52, 155, 161–62, 164–68, 173–75, 184–85
Vietcong, 73–74
Vietnam War, 72–74, 83, 116, 136
Vinokur, Roman, 101, 120n8, 131
Virilio, Paul, 8, 56, 65, 73, 86, 106, 113, 116, 126, 132–34, 142–44
Vivaldi, Antonio, 86
voguing, 156
voice, ix, xi, 5–6, 12, 14, 43, 50, 74, 82, 97–100, 105, 109, 125, 131, 137–39, 161, 165, 168, 170, 174, 181–82; of the dead, 73, 120n9, 173; of the devil, 94; of God, 33–34, 52–53; hearing voices, 138; inner voice, 16, 93, 98, 104, 133–34, 153, 181–85; scrambling, 135

Waco, 2, 13, 49–75, 75n2, 77, 115, 183
wall of sound, 111
Walonick, David, 148n1
warfare, sonic, 19n16, 40, 54, 56, 63, 65, 73–74, 86, 89, 101, 113, 130–31, 139, 178–79
Wark, McKenzie, 4
Whitecross, Mat, 79
Williams, Rhys, 62, 70
Winterbottom, Michael, 79
Wired Radio, 25–27, 41
Wolff, Harold, and Hinkle, Lawrence, 83

World War I, 23, 127, 135
World War II, 26, 40–41, 75n3, 75n4, 85, 102, 120n6, 127, 135–36
Wouterloot, Lex, 141
Wright, Steve, 129

Yerkes-Dodson law, 39–40

Zero dB, 79
Žižek, Slavoj, 56, 110
zombie, 171, 174–75

About the Authors

Dave Tompkins' first book *How to Wreck a Nice Beach: The Vocoder from World War II to Hip Hop*, was published by Stop Smiling/Melville House. He has contributed to the *Unsound : Undead* anthology (Urbanomic, 2019), and Kristen Gallerneaux's *High Static Dead Lines* (Strange Attractor, 2018), as well as *The New Yorker*, *New York Magazine*, *The Wire*, and *The Paris Review*. He is currently writing a natural history of Miami Bass for Simon & Schuster.

Dr Toby Heys is Professor of Digital Media and Head of Research for the School of Digital Arts (SODA) at Manchester Metropolitan University, England. He is also an affiliate member of Hexagram—an international network dedicated to research-creation in Media Arts, Design, Technology and Digital Culture in Montreal, Canada. He works within the AUDINT research unit producing art installations, films, software, sound recordings and books such as *Unsound: Undead* (Urbanomic, 2019).

www.ingramcontent.com/pod-product-compliance
Lightning Source LLC
Chambersburg PA
CBHW021848300426
44115CB00005B/70